CISM COURSES AND LECTURES

The series presents lecture notes, monographs, edited works and proceedings in the field of Mechanics, Engineering, Computer Science and Applied Mathematics.
Purpose of the series is to make known in the international scientific and technical community results obtained in some of the activities organized by CISM, the International Centre for Mechanical Sciences.

INTERNATIONAL CENTRE FOR MECHANICAL SCIENCES

COURSES AND LECTURES - No. 409

ENVIRONMENTAL APPLICATIONS OF MECHANICS AND COMPUTER SCIENCE

PROCEEDINGS OF CISM 30th ANNIVERSARY CONFERENCE
UDINE, May 29, 1999

EDITED BY

GIOVANNI BIANCHI
POLITECNICO DI MILANO AND CISM

Springer-Verlag Wien GmbH

This volume contains 68 illustrations

SPIN 10733045

ISBN 978-3-211-83152-6   ISBN 978-3-7091-2492-5 (eBook)
DOI 10.1007/978-3-7091-2492-5

In order to make this volume available as economically and as
rapidly as possible the authors' typescripts have been
reproduced in their original forms. This method unfortunately
has its typographical limitations but it is hoped that they in no
way distract the reader.

PREFACE

In 1999 the International Centre for Mechanical Sciences celebrates thirty years of activity. For this celebration CISM has decided to organize a series of courses and meetings on environmental problems, one of the leading subjects today of theoretical and applied research all over the world. The results obtained directly influence our daily life, particularly in applications for protection from pollution and natural hazards.

The most significant of the events planned is this May 29 Conference on "Environmental Applications of Mechanics and Computer Science", where prominent scientists in the field present significant examples of the scientific approach to large scale phenomena involved in environmental problems.

The disposal of nuclear waste is a major preoccupation in the world today. One project plans the encapsulation of transuranic waste in caverns excavated in bedded salt formations. The success of the operation requires a sure knowledge of the creep properties of rock salt. Bodner et al. investigate this aspect of the problem in their paper "Mechanics Applied to the Underground Storage of Radioactive Waste Materials".

The study of the morphology of rivers, estuaries and coasts addresses a randomly forced multiscale nonlinear process. A rational description and prediction of the behaviour and evolution of these geographic elements relies on the use of space and time scales very much larger than those needed for the analysis of the underlying motion of water and sediment. H.J. de Vriend's paper on the "Long-term Morphodynamics of Alluvial Rivers and Coasts" discusses a strategy for the development of a predictive model system that can deal with the variety of scales involved in the problem.

An effective campaign against water pollution requires an understanding of the motion of fluids and the associated mass and heat transport in the hydrosphere and atmosphere of the earth. G.H. Jirka's paper "Environmental Fluid Mechanics: its Role in Solving Problems of Pollution in Lakes" treats in particular the problems of gas transfer at the air-water interface, turbulence and pollutant transport in shallow flows, and the prediction of pollutant releases into various bodies of water.

The management of water resources to control irrigation and flooding is an ever recurrent concern in the history of man. S. Rinaldi's paper "Conflict Resolution in Water Resources Management: the Case of Lake Como" analyzes the management of Lake Como, demonstrating that it is possible to strongly reduce flooding and agricultural deficits. The method developed for the solution of the multiobjective optimization problem is general, and can be used to improve the management of any regulated lake.

The analysis of many environmental problems requires new computational techniques for the detailed study of geomechanical phenomena. B.A. Schrefler's paper "Computational Environmental Geomechanics" is based on heat and multiphase flow in the deformation of porous media, where pollutant transport mechanisms can be added. Appropriate solutions are presented for the governing equations, which are discretized by means of the finite element method in space and finite differences in time.

These papers are authoritative examples of the contribution Mechanics and Computer Science offer for the solution of the problems that assail our environment. They are the premises and the promise of a more livable world for all of us.

Giovanni Bianchi

CONTENTS

Page

Preface

Chapter 1
Long-Term Morphodynamics of Alluvial Rivers and Coasts
by H. de Vriend 1

Chapter 2
Conflict Resolution in Water Resources Management:
the Case of Lake Como
by S. Rinaldi 21

Chapter 3
Mechanics Applied to the Underground Storage
of Radioactive Waste Materials
by K.S. Chan, S.R. Bodner and D.E. Munson 31

Chapter 4
Environmental Fluid Mechanics:
Its Role in Solving Problems of Pollution in Lakes, Rivers
and Coastal Waters
by G.H. Jirka 49

Chapter 5
Computational Environmental Geomechanics
by B.A. Schrefler 99

LONG-TERM MORPHODYNAMICS OF ALLUVIAL RIVERS AND COASTS

H. de Vriend

University of Twente, Enschede, The Netherlands

Delft University of Technology, Delft, The Netherlands

Abstract. The morphology of rivers, estuaries and coasts is the result of a randomly forced multi-scale nonlinear process. Its prediction at spatial and temporal scales which are much larger than those of the underlying water and sediment motion is rather non-trivial. The paper discusses a strategy to develop a predictive model system which can handle the variety of scales.

1 Introduction

Rivers, estuaries, coastal lagoons, coasts and shallow shelf seas have one important thing in common: they shape their own boundaries (their morphology), via a complex dynamic interaction between water motion, sediment transport and bed topography changes. In general, the water motion tends to pick up bed sediment and deposit it elsewhere. This leads to changes in the bed topography, which in their turn affect the water motion. The result of this feedback process is what is usually called the "morphodynamic behaviour" of the system.

This morphodynamic behaviour takes place at a wide variety of spatial and temporal scales, from individual grain motion through to the evolution of entire systems at historical and geological time scales. It can easily be shown that this scale range extends over at least ten orders of magnitude.

The prediction of these changes at the various scale levels is not a trivial task. The present mathematical models are mostly based on descriptions of small-scale consituent processes, i.e. wave, current and sediment transport models working at time scales of minutes to hours. Practical predictions, e.g. for the impact of engineering works, are aiming at time scales of decades or more. Bridging this time-scale gap involves the question of uniqueness and deterministic predictability of the phenomena, as will be discussed hereafter. But even is the system is inherently predictable, it may be rather impractical to take a "brute-foce"approach and run these small-scale models over a huge number of time steps. In either case, we need models which are especially designed for certain scale ranges.

The present contribution gives a number of examples of morphodynamic behaviour, to illustrate its complexity and its multi-scale character. Based on what everybody can observe in nature, a cascade of scale levels is identified. The issue of predictability is discussed, and the concept of the model cascade is introduced. Finally, the need for a change in the way of prediction is discussed.

2   Free and Forced Behaviour

2.1.   General

Morphodynamic systems usually exhibit two types of behaviour. One, which we will call forced behaviour, is a direct response to the external forcing. The variations of this behaviour can be related to the variations in the forcing factors. Examples are the "breathing" of a river bed, due to discharge variations through the year, dune erosion due to a storm, and the gradual landward migration of barrier islands due to sea level rise. Other examples, associated with human-induced constraints, e.g. via engineering works, are a river's response to training works, the response of an estuary to channel deepening for navigation, the response of the Venice Lagoon to the building of jetties and shipping channels, the coastal response to groynes and jetties, the filling up of sand mining pits at sea, etc.

Figure 1. Forced response of a beach to a groyne system

The second type of behaviour globally depends on the external forcing, but the spatial and temporal variations cannot be related to similar variations in the forcing factors. This is therefore called free behaviour. Manifestations of free behaviour are ubiquitous in morphodynamic systems. The following sections give some examples.

2.2.   Free Behaviour of Rivers

The smallest-scale modes of free morphological behaviour in rivers are bed ripples, migrating undular bed features of typically one or a few decimetres high and a few metres long. They develop under moderate flow and transport conditions. If, during a flood, the discharge increases, the bedform dimensions also increase, though with a certain time lag. Ripples evolve into dunes which can be metres high and up to a hundred metres long. As the flood wanes, these large dunes slowly vanish and new small-scale ripples develop on top of them.

Via their influence on the apparent bed roughness, ripples and dunes influence the flow and the sediment transport in a river. This small-scale morphodynamic feedback system may well have large-scale effects, e.g. if the apparent bed roughness is non-uniform in space. The complexity of the interaction probably explains why bedform dynamics is seldomly taken into account in morphological predictions.

The next higher mode of free morphological behaviour in rivers is formed by the so-called alternate bars (e.g. Fukuoka, 1989). These migrating features, which are metres high and hundreds of metres long, have their crests alternately at the left and the right bank of the river. They are sometimes believed to form the onset of meandering, but they usually migrate much too fast to allow for the bank erosion which is necessary to generate a meander. Seminara and Tubino (1989) show theoretically that meandering may be influenced by the occurrence of alternate bars, but is not dependent on it. As meanders develop, the alternate bars tend to be suppressed by the meander-induced bed topography (a point bar in the inner bend, a pool in the outer bend). From a bed topography point of view, the latter is an example of forced behaviour, imposed by the curvature of the channel. At a higher scale level, however, meandering is a form of free behaviour of the channel alignment. This illustrates that the distinction between free and forced behaviour is scale-dependent.

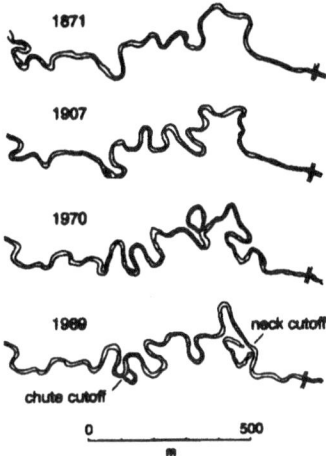

Figure 2. Meander development in the River Bollin, UK
(from: Knighton, 1998, after: Hooke and Redmond, 1992)

At the scale level of the river alignment, there are other forms of free behaviour than meandering, viz. braiding and anastomosing. There are empirical indications that these forms occur under different environmental conditions (e.g. the overall valley slope), but a theoretical foundation has not been found, so far.

At the geological scale level, channel pattern formation is generally considered to be a form of free behaviour, as well. The dynamic interaction between upland erosion and runoff results in a fractal pattern of river branches (cf. Rinaldo and Rodriguez-Iturbe, 1997). If the mouth of the river is formed by a delta, the reverse occurs: deposition at the edge of the delta causes a branched pattern of channels, which continually develop and degenerate.

2.3   Free Behaviour of Estuaries

At the lowest scale level, estuarine morphology is similar to river morphology, in that quasi-regular undular bedforms occur. There is one important difference, however, due to the changing direction of the tidal current. The asymmetry of the bedforms tends to adjust to the flow direction, but this process takes time and is probably not entirely completed before the next turn of the tide. The retarded response of the bedform asymmetry, and hence the apparent bed roughness, interacts with the tidal asymmetry, and thus with the sediment importing or exporting capacity of the system. This is another example of how small-scale phenomena can have large-scale effects.

At the next higher scale level, bed patterns may evolve in the channels, which can sometimes be attributed to the channel alignment (e.g. shallow thresholds at the transition between two consecutive bends), sometimes not. The formation of channel and shoal patterns in an estuary is another form of free behaviour, at a yet higher scale level. Finally, the estuary as such may develop large-scale meanders, as illustrated in Figure 3.

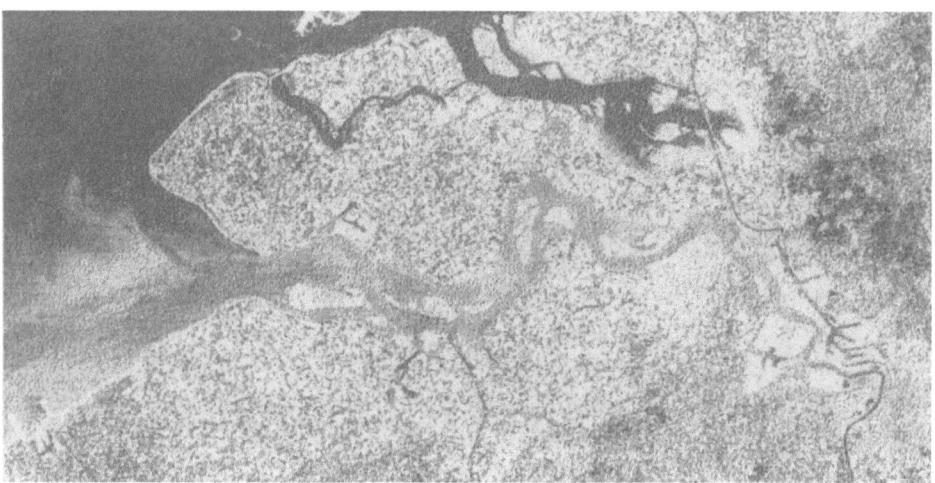

Figure 3. Overview of the Western Scheldt estuary, The Netherlands
(courtesy Rijkswaterstaat – RIKZ)

2.4   Free Behaviour of Coasts

At first sight, an uninterrupted part of coast may seem rather uniform in the alongshore direction. Upon closer inspection, however, this is not the case: the system is full of free behaviour at a wide range of scales. The smallest morphological scale is that of ripples, again, but now these are wave-dominated, instead of current or tide-dominated. The wave-dominated character manifests itself in a much more symmetrical shape, since there is no direction of preference and the flow variations are much too fast for the ripple shape to adjust to the instantaneous conditions within a wave cycle.

Figure 4. Wave-induced ripple pattern on a beach

At a higher scale level, there is the formation of ridge and runnel systems on the intertidal beach and complex nearshore bar patterns in the surf zone. During storms, these patterns tend to disappear and a longshore uniform pattern tends to be formed. This may well come under the definition of forced behaviour. During the subsequent period of moderate energy exposure, the three-dimensionality of the pattern evolves again, but there is nothing in the external forcing which can explain this. This is typically a manifestation of free behaviour. Recently developed video-based remote sensing techniques provide a relatively cheap tool to monitor surfzone bar systems on an hourly basis during many years (Lippmann and Holman, 1989; also see Aarninkhof et al., 1997). Figure 5 shows such a video-picture, which is basically a time-exposure of the surf zone; where waves break, i.e. at bars, a white zone is seen.

Figure 5. Video time-exposure of the surf zone at Duck, N.C., 09-09-1998;
white areas due to wave breaking show bar position
(courtesy Coastal Imaging Lab., Oregon State University)

The surf zone is not a zone of constant width: during storms it is considerably wider than during low-energy conditions. In the seaward part of the active zone, a large-scale bar system can be found. Globally speaking, the bars develop close to the shore and migrate offshore. During this migration process, they grow in amplitude until they reach a depth of 5-8 m (depending on the wave climate). Beyond that point, they lose height and ultimately die out. Figure 6 shows the pattern of these bars along the Holland coast.

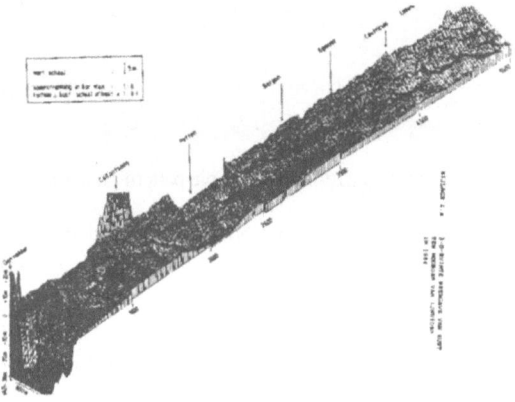

Figure 6. Pattern of breaker bars along the nortnemmost 50 km of theHolland coast
(after: De Vroeg et al., 1988)

At a yet larger scale, outside the typical nearshore environment, we find the so-called shoreface-connected ridges, large shallow features of a few metres high, with a typical wave length of 5-10 km. If they migrate, it is at a very slow pace. They occur on the Holland coast, though not everywhere, and they have also been observed on other storm-dominated coasts, such as the Atlantic coast of the USA. They can be explained from the nonlinear interaction of net currents, the associated sediment transport and the bed topography (cf. Falqués et al., 1998a).

2.5   Free Behaviour of Tidal Inlets and Lagoons

Tidal inlet systems consist of a number of dynamically coupled subsystems, viz. the basin, the inlet proper (the gorge), the outer delta and the adjacent coasts. The morphology of each of these subsystems exhibits free behaviour associated with the inlet.

The basin morphology is dominated by a system of channels and shoals, often with a fractal pattern of ever smaller channels (also see Figure 7). The shoals are considered to be an important regulating factor in the system, because they control the tidal asymmetry, an important mechanism for net sediment import or export. Thus the relatively small-scale phenomenon of channel/shoal interaction has a large-scale influence on the system as a whole.

Figure 7. Landsat image of the Wadden Sea around the Ems-Dollard estuary

Also the outer delta exhibits a variety of free behaviour. One important phenomenon, which is probably associated with the bypassing of sediment across the inlet, is the formation of a migrating bar/channel system on the delta. This system gradually migrates in the direction of the net longshore sediment transport. Figure 8 shows such a bypassing system: clearly, the sand supply to the downdrift island is associated with the point where the shoals reach the island. Since this welding of shoals is an intermittent process, the island will receive "shots" of sand, which migrate as coastline "waves" eastward along the island coast (Bakker, 1968).

Also note in this Figure what happens if man does not comply with the natural tendency of the inlet to migrate eastward. The westernmost part of the island has to be defended ever stronger in order to keep it in place.

Another mode of free behaviour is found in the lee of the outer deltas of many tidal inlets: sets of migrating bars under a sharp angle with the shore normal. These so-called sawtooth bars (Ehlers, 1988) are largely unexplored, even though they seem to indicate a significant longshore transport outside the surf zone.

At historical and geological scales, barrier island coasts are extremely interesting. Inlets tend to migrate more or less systematically within certain geological constraints, and the roll-over of barrier islands is one of the mechanisms of adjustment to sealevel rise (cf. Cowell et al., 1997). Although this roll-over, as such, should probably defined as forced behaviour, the formation of barrier islands is definitely a manifestation of free behaviour.

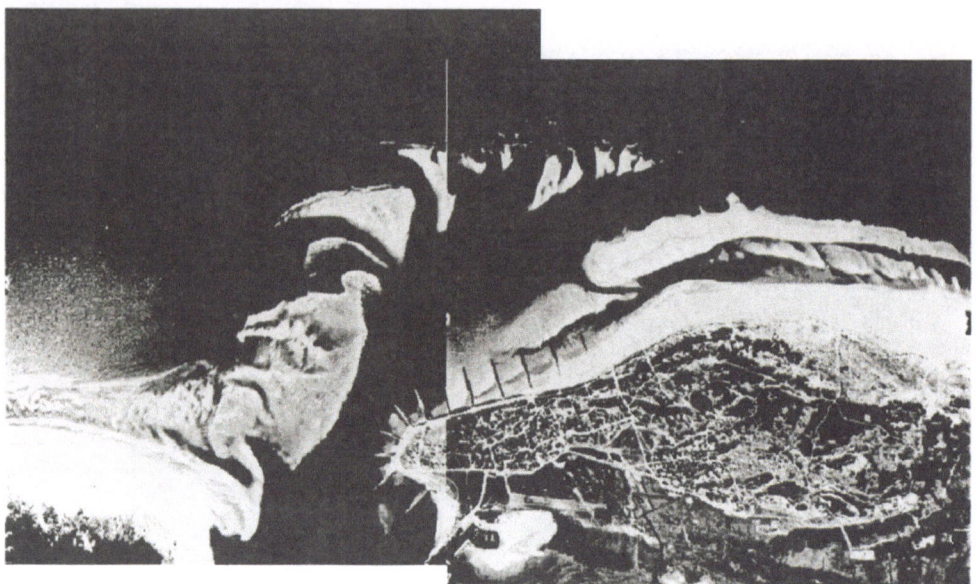

Figure 8. Shoal pattern on the outer delta between Norderney and Baltrum, East Frisian Wadden Sea
(from: J. Ehlers, *Morphodynamics of the Wadden Sea,* Balkema, Rotterdam, 1988)

2.6   Free Behaviour of Shelf Seas

Even if we disregard small-scale bedforms ranging from ripples to megaripples, the seabed is
by far not as flat and inert as one might expect. Especially if tidal currents and stormwaves are
strong enopugh to mobilize the sediment, various forms of free behaviour are found. The largest
scale concerns systems of long-crested ("linear") sandbanks, with a height up to 20 m and a wave
length of typically 10 km. They occur in the North Sea, but also in other shallow shelf seas. There
has been a dispute about whether these are active features, or relicts from former times. Analyses
based on the assumption that they are active morphodynamic features (Huthnance, 1982; Hulscher,
1996) have revealed that the time scale of their evolution amounts centuries to millennia, which
makes the dispute somewhat academic. Yet, sandbanks are practically interesting features, if it
were only because they can be places where hydrocarbons have piled up.

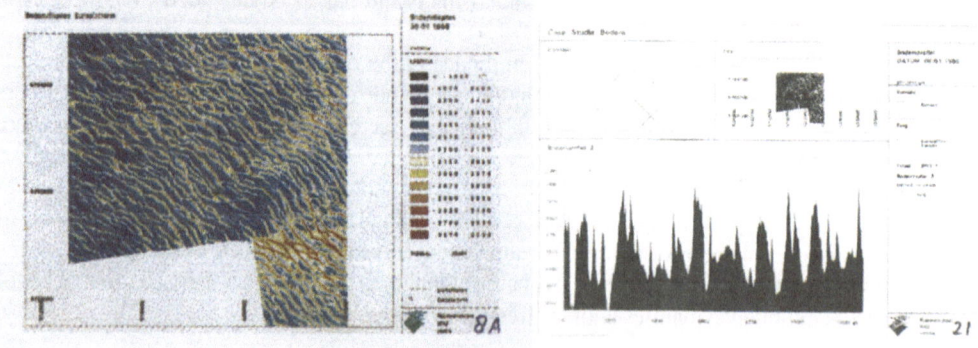

Figure 9. Observed sandwave pattern near the Europlatform, off Rotterdam
(from: Van Goor and Andorka, 1996)

Practically even more interesting are sandwaves, significantly smaller features of a few hundres of metres wavelength and a height up to some 10 metres. Figure 9 shows a snapshot of a sandwave field of 10 by 10 km off Rotterdam, where sandwave heights of 8 m are found at a mean water depth of 30 m. Like most free morphological patterns, it looks quasi-regular, with a predominant length scale and a predominant orientation. Sandwaves are much more mobile than sandbanks and may cause trouble to navigation channels and piplelines. Moreover, their potential of segregating sand fractions between their crests and troughs makes them interesting for sand mining.

3   Scale Cascade

In each of the systems considered in the foregoing, a series of scale levels can be distinguished. Assuming that to some extent these scale levels can be considered separately, they form a sort of cascade (Figure10). At each step of the cascade, morhodynamic processes have to be considered in mutual interaction. In the following subsections, we will further specify the cascade for various types of morphodynamic systems.

3.1   Scale Cascade for River Morphology

The micro-scale level in river mophology represents the small-scale bedforms (ripples and dunes), but also the vertical segregation of sediment fractions (bed armouring, etc.). The meso-scale level is that of alternate bars and cross-sectional profile evolution (e.g. pointbar/pool combinations in bends). At the macro-scale level we have meandering, braiding and anastomosing, but also the longitudinal profile evolution of river reaches, e.g. in response to training works or sand mining. Channel pattern formation at the scale of the river basin constitutes the mega-scale.

3.2   Scale Cascade for Estuarine Morphology

The micro-scale level in estuarine morphology is that of the small-scale bedforms (ripples, megaripples), but also of the small-scale segregation of sand and mud, e.g. due to a concentration of filter feeders. At the meso-scale level we find channel profile evolution, threshold formation,

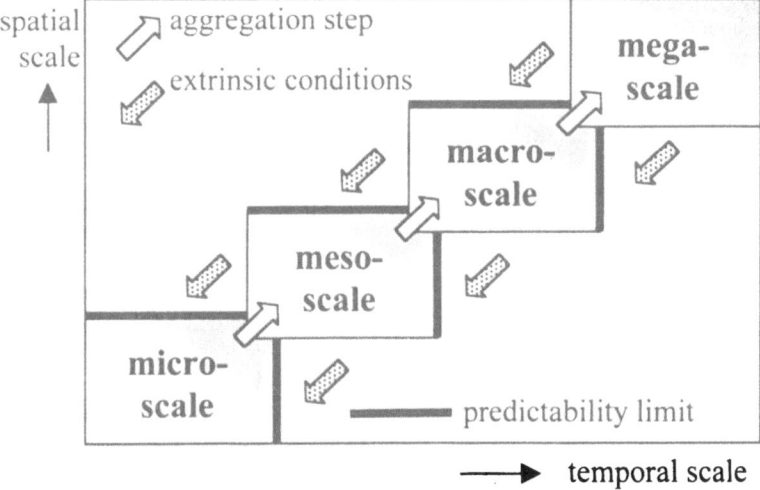

Figure 10. Scale cascade

secondary channel formation (shortcut channels), etc. The macro-scale level concerns the primary channel and shoal system, including meandering, intertidal flat processes, marsh formation, etc. At the mega-scale level, we have the evolution of the estuary as awhole, in connection to its outer delta and possible neighbouring estuaries (cf. the Dutch Delta region, where before the Deltaworks a number of estuaries used to share one outer delta).

3.3   Scale Cascade for Coastal Morphology

In coastal morphology, the scale cascade is built up as follows:
-    micro-scale: small-scale bedforms (wave ripples) and vertical sediment segregation processes
-    meso-scale: beach, nearshore and breaker bar systems, coastlines waves (as footprints of nearshore bar systems), horizontal segregation of sediment, dune erosion, beach response to groyne fields, profile response to nourishments, etc.
-    macro-scale: longshore variations of bar –system properties (cf. Wijnberg and Terwindt, 1995), coastline response to jetties, detached breakwaters and major engineering works, shoreface profile steepening, etc.
-    mega-scale: coastal response to sealevel rise, evolution of coastal cells due tovariations of sediment input, etc.

3.4   Scale Cascade for Tidal Inlet Morphology

The scale cascade for tidal inlets can be built up as follows:
-    micro-scale: small-scale bedforms, vertical sediment segregation processes
-    meso-scale: bar/channel formation at the outer delta, sawtooth bar systems in the lee of the delta, channel/shoal interaction inside the basin, formation of marshes, mussel banks, etc.
-    macro-scale: interaction of overall properties of the basin (e.g. volume below mean sealevel, total intertidal area, average flats level), the inlet proper (cross-sectional area), the outer delta (e.g. volume, protrusion) and the adjacent coasts, response to medium-scale interferences (NL: closure of the Lauwerszee), response to large engineering works (e.g. jetties and large navigation channels in the Venice Lagoon), response to regional subsidence due to hydrocarbon mining, etc.
-    mega-scale: behaviour of barrier-island coast as a whole, large-scale interaction between neighbouring inlet systems, response to major interferences (NL: closure of the Zuiderzee), response to sealevel rise, etc.

3.5   Scale Cascade for Shelf Sea Morphology

In shelf seas, the micro-scale phenomena are primarily small-scale bedforms, but also the effects of fishing (some parts of the North Sea bottom are ploughed almost dayly by fishermen). Sand waves are a typical meso-scale phenomenon, and so are the responses to navigation channels and medium-scale sand mining (e.g. for shore nourishments). At the macro-scale level we find sandbanks, but also the response to large-scale sand mining (e.g. for major land reclamation projects), artificial offshore islands, etc. At the mega-scale level, we have the evolution of the shelf sea basin as a whole, under the influence of sealevel rise, tectonics, etc.

4   Predictability

The issue of inherent unpredictability used to be hardly explored in morphodynamic modelling. To physical scientists, this may seem surprising, because the system is definitely nonlinear, it undergoes a continuous input of energy which is internally dissipated, it exhibits many types of free behaviour, and it is randomly forced. This combination is a good reason to suspect the possibility of deterministically unpredictable behaviour beyond certain limits.

In recent years, significant progress has been made in linear and nonlinear analyses of morphodynamic systems, though usually with highly idealised models. At the University of Genoa, a group of mathematically oriented engineers and scientists has worked since more than fifteen years on nonlinear analyses of river morphology (alternate bars and bends; see Seminara and Tubino, 1989), seabed morphology (e.g. Vittori and Blondeaux, 1992), beach cusps (e.g. Vittori, De Swart and Blondeaux, 1998), sandwaves, etc. In the late eighties, a group of mathematicians and oceanographers at the University of Utrecht took up a similar line of research, dealing with alternate bars (Schielen et al., 1992), sandbanks and sandwaves at the seabed (Hulscher, 1996), tidal inlet dynamics (Schuttelaars and De Swart, 1998), estuaries, etc. Especially the latter studies have given clear indications of multiple attractors and limited predictability. In recent years, other groups have joined in, such as UPC Barcelona (e.g. Falqués et al., 1998a), the University of Twente (e.g. Hulscher and Roelvink, 1998), HR Wallingford (cf. Falqués et al., 1998b) and the University of Warwick (Komarova and Hulscher, 1999).

This formal evidence definitely needs further pursuing, including confrontations with data and more sophisticated models. On top of this, we should also look for empirical evidence, be it circumstantial, of the possible occurrence of inherent predictability limits. This includes a fractal structure of the system's state, more than one possible state for a given input (multiple attractors), vacillation ("hesitation") between directions of evolution, jumping between different states, or even deterministic chaos.

Fractal structure. The state of a dynamic system has a fractal structure if the same pattern occurs at a discrete series of scale levels. In the case of morphodynamic systems, we therefore have to be aware of the possibility of a fractal structure if the same topographical patterns show up at different scale levels. As we have seen in the foregoing, bedform patterns of a similar nature do occur at a series of scale levels (ripples, megaripples, sandwaves, sandbanks), and indeed without forming a continuous spectrum. The patterns show clear signs of nonlinearity, such as crest bifurcations, crest splitting ("eyes") and quasi-regularity. Linear systems can only yield (combinations of) regular free modes.

Southgate and Möller (1998) investigated the fractal properties of a beach profile near the USACE Field Research Facility at Duck, N.C., which has been monitored on a weekly basis since almost 20 years now. They found that the system exhibits fractal behaviour and is uncorrelated to the variations of the incoming wave energy within a certain time window. This window varies across the profile, from 1-20 months for the beach and dune area, via 30-40 months between the inner and the outer bar zone, back to 1-12 months for the upper shoreface, i.e. the zone offshore of the outer bar. Outside this window, the profile behaviour is well-correlated to the forcing factors, but inside it self-organisation and limited predictability are to be expected.

Vacillation and jumping. So far, little evidence of vacillation has been reported, but this may just as well be due to the fact, that a clear picture of the attractor states is still missing. It is rather difficult to assess whether a system is "hesistating" between two attractor basins if we are not even aware of their existence.

Plant and Holman (1997) analysed the location of the crest of the outer bar at Duck. Their analysis shows a number of very sharp transitions, which cannot be attributed to similar variations in the forcing. It also indicates that the local bar behaviour is strongly nonlinear and cannot always be related to variations in the input. On the other hand, Plant et al. (1998) show that, after averaging over a certain distance alongshore, the bar crest position is well-correlated to the wave input and can be described by a relatively simple model. Apparently, the strongest nonlinearity occurs at a finite longshore length scale.

As more video images become available and get analysed, the picture of nearshore bar dynamics is likely to become clearer. The analysis of video images of the nearshore bar patterns observed at Duck, N.C. showed, after the first few years, a rather consistent pattern of delayed response to the wave energy input: during storms, a rapid stretching of the outer bar occurred, and during the subsequent periods of less energy exposure, the pattern tended to become gradually more three-dimensional, up to a well-developed crescentic bar system (Lippman and Holman, 1990). Upon the impact of a few extreme storms, around 1990, however, the behaviour changed drastically: it became much more irregular and thus it has remained until now, even though the input conditions have not changed essentially since the beginning of the observation period (Lippmann et al., 1993). This could be a case of jumping between two attractors. Apparently, the storms were just enough to put the system over the edge of its "attractor basin".

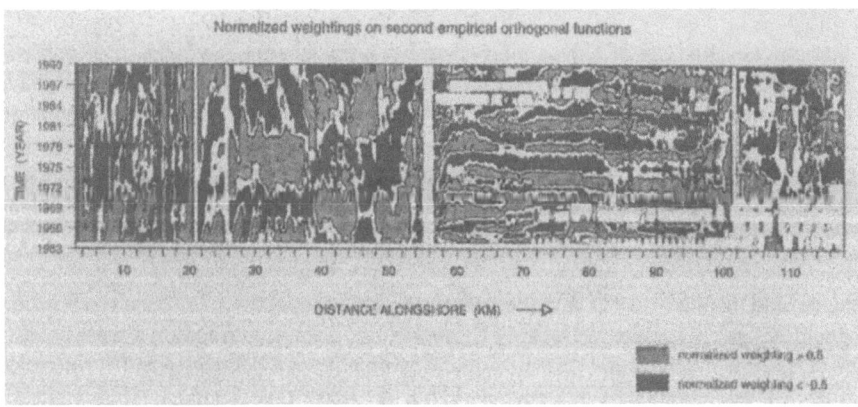

Figure 11. Provinces in breaker bar behaviour along the Holland coast, as revealed by an EOF analysis; red and blue in this time-space plot indicate the phase of the bed waves; difference in pattern indicate "provinces" (courtesy: K. Wijnberg, University of Utrecht)

Another indication of jumping between states, but now in space, stems from an EOF-analysis of the JARKUS-data for the Holland coast (Wijnberg and Terwindt, 1995). This extensive dataset covers more than 25 yearly records of the cross-shore profile every 250 m along the entire coast of the Netherlands. The analysis revealed the existence of distinct "provinces" in the interannual behaviour of the nearshore bars (Figure 11). They differ especially in the return period of the outer bar, and in its migration speed. They are separated by surprisingly narrow transition zones, sometimes consisting of large engineering works (e.g. the entrance of IJmuiden harbour), sometimes without any obvious trigger.

The occurrence of these "provinces", which can by no means be related to longshore variations in wave climate, tidal motion or sediment properties, might be interpreted as if bar behaviour be a matter of extrapolation from the past, such that accidental differences determine where the system goes (cf. the development of a language in two isolated areas). This would mean that bar formation would be an initial value problem, which seems rather unlikely, because of the ubiquitous spatial interactions. Another interpretation might be that the co-existence of different lines of evolution is associated with different attractors.

5   Model Cascade

Even if there are no inherent predictability limits within the range of practically relevant scale levels, there can be many kinds of other limitations to the predictive capabilities of a model. One of them is the pile-up of numerical erros after many small time steps, another one the accumulation of inaccuracies due to model simplifications. Capobianco (1998) defines the so-called "window of predictability", i.e. the area in the space-time domain where a certain predictive model is able to produce reliable results. This area is bounded by various types of predictability limits, among which the inherent one (which is not necessarily critical).

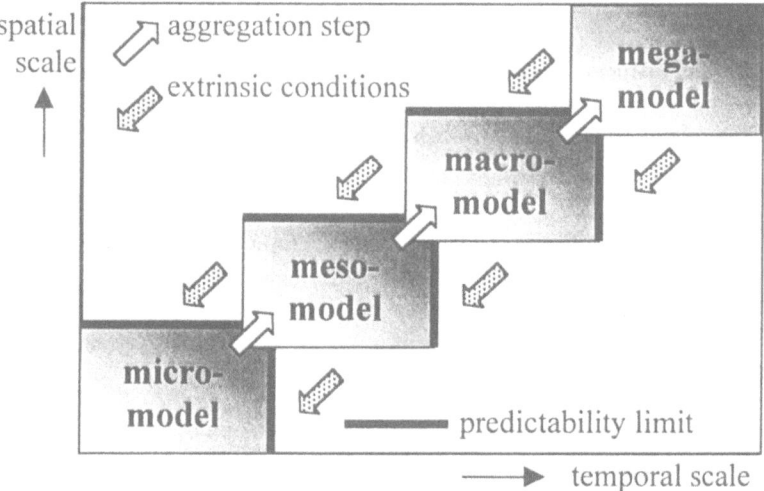

Figure 12. Model cascade

The window of predictability of a certain model may well be considerably smaller than the total range of scales which we wish to cover. In that case, we need a combination of different models of which the windows cover the entire range (Figure 12). Each model may be derived from the one at the next lower scale level via some form of aggregation or extrapolation, or it may be designed separately for the scale level it is supposed to work at. In hydrodynamic and morphodynamic modelling we find examples of either possibility.

Model aggregation. One example of formal model aggregation is the modelling of turbulent mean flow. Models of this type are based on some form of the Reynolds equations, which are the Navier Stokes equations formally averaged over the turbulent motion. Upon averaging, the nonlinearity of these equations yields residual terms, which have to be linked to mean flow properties via some turbulence model.

Models of wave-driven currents are based on the wave-, turbulence- and sometimes also depth-averaged flow equations. The wave-induced residual terms associated with the nonlinear terms in the equations and the boundary conditions are modelled using some wave theory. Thus Longuet-Higgins and Stewart (1962) arrived at the wave-induced radiation stress.

Other ways of aggregation are the tabulation or the parameterisation of integral properties of small-scale model results. The STP sediment transport model, for instance (Deigaard et al., 1986), is based on a detailed description of the intra-wave water and sediment motion. Its application in multi-dimensional morphodynamic models, however, requires a much faster determination of the transport in each grid point and at each time step. Therefore, the magnitude and the direction of the wave-averaged transport are tabulated against a number of relevant hydrodynamic parameters and sediment properties.

A model can also be aggregated by fitting a larger-scale model to it. The aggregation associated with the fitting process provides a shift of the predictability horizon, at the expense of a loss of detail. One example is the fitting of a diffusion-type coastal profile model to the results of a process-oriented profile model based on detailed descriptions of wave, current and sediment transport processes (cf. De Vriend et al., 1993a). All complexity of the detailed model is stored in the diffusion coefficient, and the remaining large-scale model is reduced to the essence: a slope-dependent cross-shore transport model combined with a sediment balance. The diffusion coefficient, however, has a much smaller number of degrees of freedom than the original model, so information is lost.

Model extrapolation. Assuming the existence of a static equilibrium state, morphological models can also be extrapolated in time to their equilibrium state. In principle, this state does not have to be realistic, in that it may never be reached (before that time, the external conditions have long changed). Yet, the equilibrium state tells something about the situation which the system tends to and may help to speed up the time-stepping procedure in long-term models. An example of such an equilibrium-state model is described in De Vriend et al., 1993b). It concerns a 2-D horizontal model, in which the flow pattern in the horizontal plane is assumed to be fixed, i.e. the flux of water is invariant in every point of the model domain. Thus flow continuity is guaranteed. The water depth and the flow velocity are adjusted in such a way, that also the conservation of sediment is satisfied, under the constraint that the bed topography is not changing. Using the resulting equilibrium bed topography as a predictor, much larger time steps can be taken than in a straightforward simulation model. One can also use this model to estimate the equilibrium bed topography by some iterative procedure, updating alternately the flow pattern and the corresponding equilibrium bed topography. This technique has been utilised in the long-term prediction of the impact of sand mining schemes for the extension of Rotterdam Harbour (Walstra et al., 1997).

One may have theoretical objections against this approach, claiming that a steady equilibrium does not necessarily exist, or that the steady input which is required for a steady equilibrium does not occur in nature. Steadiness, however, has a time-scale dimension here: no variations may occur at time scales of the order of magnitude of the time needed for the system to reach its equilibrium state. Variation at essentially shorter time scales, however, are allowable.

Semi-empirical modelling. Instead of using smaller-scale models as a source of information, one can also use measured data. In that case, a model concept, often in terms of certain nondimensional combinations of parameters, is fitted to available data. Sediment transport formulae are a good example of this technique. Instead of going through the complexity of randomly forced nonlinear multi-grain interactions, the net sediment flux is expressed in terms of mean flow properties by fitting a curve in the nondimensional parameter space to the data. Another example are the empirical relationships for equilibrium state properties of tidal morphodynamic systems, such as O'Brien's (1969) relationship for the channel cross-sectional area, and Walton and Adams's (1976) relationship for the volume of the outer delta, both expressed in term of the tidal prism. These relationships can be embedded in mathematical models, which at least make sure that the sediment balance is satisfied and maybe introduce some additional physics (e.g. De Vriend et al., 1994). Since the empirical relationships can be considered as filtered representations of a large amount of data , these combined models can be considered as aggregated with respect to the short-term variability of these systems.

In summary, there is a range of existing and more or less widely applied aggregation techniques which are presently in use. Collaboration with other fields of science and further research may lead to other methods which are as yet unknown in morphodynamic modelling.

6   Prediction Methodology

In the foregoing we have shown that the prediction of the large-scale morphological behaviour of rivers, estuaries and coasts is something totally different from hindcasting, and that it involves much more uncertainty than one might have expected at first sight. This is due to uncertainties in the input, in the inherent system behaviour, in the model concept, in the numerical solution technique, in the parameter setting, etc. This situation calls for abandoning the deterministic approach used so far and moving over to probabilistic prediction methods. Attempts so far with relatively simple models have revealed considerable ranges of variation, even though the input time series were statistically equivalent (Vrijling and Meijer, 1992; Southgate, 1995). Also, rather suspect result were obtained by models which had gloriously passed deterministic hindcast tests (Aarninkhof et al., 1998). Therefore, probabilistic modelling of morphodynamic systems is definitely an area of further research.

By making good use of data and idealized model analyses, it may also be possible to trace multiple attractors, if they exist. Yet, this is likely to be a difficult and vulnerable process, unless we manage to develop efficient searching algorithms aimed at finding these attractors.

If we succeed in reaching these goals, we will be able to give an indication of the range of possible behaviour according to the model. This will be a much more useful result than just stating that the bed level in point x at time t will be equal to z, withouth giving a confidence band or an indication of possible alternative developments.

On top of this, we need indicators of the skill of the model. This skill will depend on the model concept, the parameter setting, the phenomena to be predicted, etc. Moreover, skill is not a static property, but a function of the time span covered. Presumably, more than one indicator will be needed in order to cover all relevant skill variations.

The implications of this change are rather far-reaching. First of all, we have to think in terms of probability and risk, rather than in terms of deterministic bed evolutions as a function of place and time. This will turn out to be a good investment, since the end users of our predictions think in similar terms (risk, options, manoeuvering space, flexibility, resilience, etc.). The translation of morphological predictions into these terms deserves much more attention than it has received, so far. Close collaboration with end users is an absolute necessity here.

Another implication of moving over to probabilistic predictions is that we will have to speed up our models drastically, since a probabilistic approach usually requires many more runs than a deterministic one. The reliability of the resulting statistical estimators has to be monitored throughout the running process, because their evolution with the number of runs is not necessarily smooth.

Finally, we have to spend more research effort to identifying the essential properties of natural time series that drive morphodynamic systems. This must enable us to synthesize the large sets of possible future time series which are needed to drive a probabilisitc prediction model.

The introduction of more or less objective measures of skill makes it possible to compare models from different sources. This may seem threatening to their owners, but in the long run it will turn out to be a blessing to the further development of these models. Moreover, each of the major players in this league has his models tested against others and against data, so surprises are not to be expected. Experience so far (e.g. Roelvink and Brøker, 1993; De Vriend et al., 1993b; Nicholson et al., 1998) has shown that morphodynamic model intercomparisons always yield a rather scattered picture, without a distinct "winner". This shows that the advantages of using objective measures of skill (clearer picture of reliability, stronger feedback to research) are likely to be predominant.

7   Conclusion

The conclusion from these considerations is, that natural morphodynamic systems are randomly forced nonlinear dissipative systems with a continuous energy input, that they exhibit various modes of free behaviour and transfer energy up and down the scale cascade. They are therefore susceptible to deterministically unpredictable behaviour and need to be analysed at that point.

One implication may be, that it is inherently impossible to run models based on small-scale process descriptions through to much larger scales. But, even if that turns out not to be the case within the practically relevant scale ranges, it is probably necessary for efficiency and reliability reasons to split that scale range into a number of distinct levels, at which different models are apllied. Such a model should be linked to that at the next lower scale levels via some form of aggregation, and to that at the next higher scale level via a transfer of boundary conditions and other constraints.

In order to cope with the uncertainties involved in large-scale morphodynamic modelling, predictive models should be used in a probabilistic mode, in combination with quatitative and objective measures of skill. Research and development efforts are needed in order to speed up the models, to better specify the properties of synthetic input time series (statistical equivalence is probably not enough), to learn how to think about morphology in probabilistic terms, and to simply build up experience in this domain.

8   Acknowledgement

Many of the ideas laid down in this contribution have been developed in the EU-sponsored PACE-project, in the framework of the MAST-III programme, contract no. MAS3-CT95-0002. The author is indebted to his colleagues in this project, for discussing and challenging these ideas with utmost patience.

9   References

Aarninkhof, S.G.J., Janssen, P.C. and Plant, N.G., 1997. Quantitative estimations of bar dynamics from video images. In: E.B. Thornton (ed.): *Coastal Dynamics '97*, ASCE, New York, p. 365-374.

Aarninkhof, S.G.J., Hinton, C. and Wijnberg, K.M., 1998. On the predictability of nearshore bar behaviour. In: B.L. Edge (Editor): *Coastal Engineering 1998*, ASCE, New York (in press).

Bakker, W.T., 1968. A mathematical theory about sand waves and its application on the Dutch Wadden isle of Vlieland. *Shore and Beach*, 36(2): 4-14.

Capobianco M., 1998. Predictability for Long-term Coastal Evolution - Handling the Limiting Factors. Keynote address to *Third European Marine Science and Technology Conference* - Scientific Colloquium: Predictability for Long-term Coastal Evolution - Is it Feasible?, Lisbon, Portugal

Cowell, P.J., Hanslow, D.J. and Meleo, J.F., 1997. The shoreface. In: A.D. Short (ed.): *Beaches and shoreface morphodynamics*. Wiley, New York.

Deigaard, R., Fredsøe, J. and Brøker-Hedegaard, I., 1986. Suspended sediment in the surf zone. *Journal of Waterway, Port, Coastal, Ocean Engineering*, 112(1): 115-128.

De Vriend, H.J., Bakker, W.T. and Bilse, D.P., 1994. A morphological behaviour model for the outer delta of mixed-energy tidal inlets. *Coastal Engineering*, 23(3&4): 305-327.

De Vriend, H.J., Capobianco, M., Chesher, T., De Swart, H.E., Latteux, B. and Stive, M.J.F., 1993. Approaches to long-term modelling of coastal morphology: a review. *Coastal Engineering*, 21(1-3): 225-269.

De Vriend, H.J., Zyserman, J., Nicholson, J., Péchon, Ph., Roelvink, J.A. and Southgate, H.N., 1993a. Medium-term 2-DH coastal area modelling. *Coastal Engineering*, 21(1-3): 193-224.

De Vroeg, J.H., Smit, E.S.P. and Bakker, W.T., 1988. Coastal Genesis. In: B.L. Edge (ed.), *Coastal Engineering 1988 Proceedings*, ASCE, New York, p. 2825-2839.

Ehlers, J., 1988. *The Morphodynamics of the Wadden Sea*. Balkema, Rotterdam, 397 pp.

Falqués, A., Calvete, D. and De Swart. H.E., 1998a. Shoreface-connected sand ridges and long term dynamical coupling between topography and currents. To be published in *Journal of Fluid Mechanics*.

Falqués, A., Calvete, D., De Swart, H.E. and Dodd, N., 1998, Shoreface connected ridges: a simple mechanism for their initiation and evolution. In: B.L. Edge (ed.): *Coastal Engineering 1998*. ASCE, New York (in press).

Fukuoka, S., 1989. Finite amplitude development of alternate bars. In: G. Parker and S. Ikeda (eds.): *River Meandering*. AGU Water Resources Monograph 12, p. 237-265.

Hooke, J.M. and Redmond, C.E., 1992. Causes and nature of river planform change. In: P. Billi, R.D. Hey, C.R. Thorne and P. Tacconi (eds.): *Dynamics of Gravel-bed Rivers*. Wiley, Chichester, p. 557-571.

Hulscher, S.J.M.H., 1996. Tidally induced large-scale regular bed form patterns in a three-dimensional shallow water model. *Journal of Geophysical Research*, 101(C9): 20,727-20744.

Hulscher, S.J.M.H. and Roelvink, J.A., 1998. Large-scale seabed features in the North Sea: comparison between theory and observations. In: K.P. Holz et al. (eds.): *Advances in Hydro-Science and Engineering.* University of Mississippi, Center for Computational Science and Engineering (full paper on CD-ROM).

Huthnance, J., 1982. On one mechanism forming linear sand banks. *Estuarine, Coastal and Shelf Science*, 14, p.79-99.

Knighton, D., 1998. *Fluvial Forms and Processes: a New Perspective.* Arnold, London, 383 pp.

Komarova, N.L. and Hulscher, S.J.M.H., 1999. Linear instability mechanisms for sandwave formation. Submitted for publication.

Lippmann, T.C. and Holman, R.A., 1989. Quantification of sand bar morphology: a video technique based on wave dissipation. *Journal of Geophysical Research*, 94(C1): 995-1011.

Longuet-Higgins, M.S. and Stewart, R.W., 1964. Radiation stresses in water waves; a physical discussion with applications. *Deep Sea Research.*, 11: 529-562.

Nicholson, J., Brøker, I., Roelvink, J.A., Price, D., Tanguy, J.M. and Moreno, L., 1997. Intercomparison of coastal area morphodynamic models. *Coastal Engineering*, 31(1-4): 97-123.

O'Brien, M.P., 1969. Equilibrium flow areas of inlets on sandy coasts. Proc. ASCE, *Journal of Waterways, Harbours and Coastal Engeering.*, 15(WW1): 43-52.

Plant, N.G. and Holman, R.A., 1997. Strange kinematics of sandbars. In: E.B. Thornton (ed.): *Coastal Dynamics '97*, ASCE, New York, p. 355-363.

Plant, N.G., Holman, R.A. and Freilich, M.H., 1998. Lessons learned from a simple model for sand bar behavior. In: *EOS, Transactions of the American Geophysical Union*, Fall Meeting 1998, p. 450.

Rinaldo, A. and Rodriguez-Iturbe, I., 1997. Dynamics of self-organization and the fluvial landscape: a nonreductionist perspective. In: F.M. Holly and A. Alsaffar: *Managing Water: Coping with Scarcity and Abundance.* (Proceeding 27th IAHR Congress), ASCE, New York, p. 34-39.

Roelvink, J.A. and Brøker, I., 1993. Cross-shore profile models. *Coastal Engineering*, 21(1-3): 163-191.

Roelvink and Walstra, 1998 ?????

Schielen, R., Doelman, A. and De Swart, H.E., 1992. On the nonlinear dynamics of free bars in straight channels. *Journal of Fluid Mechanics*, 252: 325-356.

Schuttelaars, H.M. and De Swart, H.E., 1998. Formation of channels and shoals in a short tidal embayment. To be published in *Journal of Fluid Mechanics.*

Seminara, G. and Tubino, M., 1989. Alternate bars and meandering: free, forced and mixed interactions. In: G. Parker and S. Ikeda (eds.): *River Meandering.* AGU Water Resources Monograph 12, p. 267-320.

Southgate, H.N., 1995. The effects of wave chronology on medium and long term coastal morphology. *Coastal Engineering*, 26(3/4): 251-270.

Southgate, H.N. and Möller, I., 1998. Fractal properties of beach profiel evolution at Duck, North Carolina. Submitted for publication.

Van Goor, S. and Andorka, Gal, J.H., 1996. *Bodem.* Rijkswaterstaat, document RIKZ/OS-96.109X, 28 pp (in Dutch).

Vittori, G. and Blondeaux, P., 1992. Sand ripples under sea waves, Part 3: brick pattern ripple formation. *Journal of Fluid Mechanics*, 239: 23-45.

Vittori, G., De Swart, H.E. and Blondeaux, P., 1998. Crescentic forms in the nearshore region. To be published in *Journal of Fluid Mechanics.*

Vrijling, J.K. and Meijer, G.J., 1992. Probabilistic coastline position computations. *Coastal Engineering*, 17(1/2):1-23.

Walstra, D.J., Reniers, A., Roelvink, J.A., Wang, Z.B., Steetzel, H.J., Aarninkhof, S.G.J., Van Holland, G. and Stive, M.J.F., 1997. Large-scale Long-term Effects of Maasvlakte-2 and Related Sand Mining Schemes. Alkyon/Delft Hydraulics, Report Z2225/A194, 90 pp. (in Dutch).

Walton, T.L. and Adams, W.D., 1976. Capacity of inlet outer bars to stor sand. In: *Proc. 15th ICCE, Honolulu.* ASCE, New York, 1919-1937.

Wijnberg, K.M. and Terwindt, J.H.J., 1995. Extracting decadal morphological behaviour from high-resolution, long-term bathymetric surveys along the Holland coast using eigenfunction analysis. *Marine Geology,* 126: 301-330.

CONFLICT RESOLUTION IN WATER RESOURCES MANAGEMENT
THE CASE OF LAKE COMO

S. Rinaldi
Politecnico di Milano, Milan, Italy

Abstract. The author presents the results of his analysis of some proposals for improving the management of Lake Como. In particular, he demonstrates that it is possible to reduce flooding and agricultural deficits to less than half their historical values. To do so he first had to solve a multiobjective optimization problem, and then determine which of the efficient solutions seemed to achieve the best compromise among the conflicting interests. The method used for the analysis is general, and can be employed to improve the management of any regulated lake. It has the salient characteristic of exploiting the management's experience to the full, and of proposing operating policies relatively similar to those used in the past.

1 Introduction

We summarize here the most significant results obtained in a study of the optimal real time operation of Lake Como (for a more detailed description, see Garofalo *et al.* (1979, 1980) and Guariso *et al.* (1981, 1986)).

Many of the results discussed here can be applied to other lakes. The proposals that emerge are also of considerable economic interest as they solve, at least in part, the problem of floods in the city of Como. The study itself represents an innovation from the methodological point of view: in analyzing the problem of the real time operation of a lake, it attaches great importance to the experience of the manager. This approach makes it possible to avoid proposals that are too abstract, such as the solutions often advanced by optimization studies structured from scratch. In fact, it is precisely because our point of departure was the experience of the manager that we were able to achieve the results illustrated here.

The paper is organized as follows. The next section briefly illustrates the problem of the real time operation of the lake, and in particular its conflicting objectives, together with the information-gathering and decision-making structure that allows the day by day assessment of the amount of water to release from the regulation dam. In the third section the "historical" operating rule used implicitly by the manager in the period (1969-1978) is identified. The fourth section describes how, starting from this rule (which sums up all the experience of the manager), a new operating rule has been determined that would result , on the average, in fewer days of flooding per year, a lower agricultural deficit, and increased power production. The final section examines the possibility of obtaining further improvements through the appropriate use of information concerning the watershed (rainfall, temperature, snow pack, etc.) transmitted to the manager systematically and in real time.

2 Description of the problem

The real time operation of Lake Como began in 1944 after the construction of the Olginate regulation dam. The objective was to satisfy the contrasting demands of agricultural users and hydroelectric plants, and protect the inhabitants of lakeside localities (the city of Como in particular) from floods. These contrasting needs can be summed up in the fact that while users below the lake are interested in filling it to ensure valid water reserves against possible periods of drought, the lakeside population is interested in maintaining the level of the lake rather low, to lessen the possibility of serious flooding. Moreover, agricultural and hydroelectric enterprises demand widely varying modalities of water supply, the former requiring pronounced seasonal variations, and the latter, a supply as constant as possible.

The users downstream of the regulation dam include six major agricultural areas, served by as many irrigation canals, and seven hydroelectric plants (see Figure 1). The irrigation canals are characterized by a nominal daily water demand w_i (t) $(i = 1, ..., 6; t = 1, ..., 365)$, varying during the year in function of the state of the crops, while the power stations are characterized by the maximum capacity of the turbines w_i $(i = 7, ..., 13)$ and by the available elevation, which is, obviously, constant in time.

This complex system of distribution has been represented in a simulation model that takes into account the operating modes used to allocate the available resource. The model makes it possible to calculate the quantity of water $q_i(t)$ $(i = 1, ..., 13)$ supplied to the i-th user on day t starting from the flow $r(t)$ released from the regulation dam.

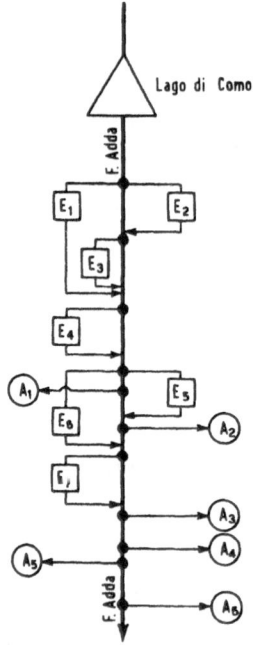

Figure 1. Distribution network below the regulation dam (A_i =agricultural areas;
E_i =hydroelectric power plants)

Thus for each user i, a daily deficit d_i (t) can be computed:

$$d_i(t) = \max\{0; (w_i(t) - q_i(t))\}$$

These deficits are the most appropriate elementary indicators on which to base a global evaluation of the objectives of the real time operation of the dam.

The evaluation of the damage to agriculture is not an easy task. Crop failure depends upon the development of the plants over the entire season, and is, influenced by the amount of water supplied, as well as by the weather conditions. Moreover, crop failure is also a function of the irrigation technique of each agricultural user. Lacking precise information about the way the individual users operate and about the growth dynamics of the various crops, the total deficit of water supply in the growing season was selected as the annual indicator of crop failure. Since this indicator, from now on called agricultural deficit, is a stochastic variable, we have assumed that the damage suffered by agriculture on the average can be indirectly evaluated by means of the average annual *agricultural deficit*, measured in millions of cubic meters.

As for the deficit in the production of hydroelectric power, the plants on the Adda river below the lake are all of the same type, with very similar efficiencies. Moreover, the number of machines in each plant is relatively high, so that the efficiency does not vary with the flow passing through the turbines. In order to avoid investigating on the operational criteria of each individual plant it seemed opportune to express the hydroelectric objective of the real time operation only in terms of the total hydraulic power made available to the plants. More precisely, the indicator used was the annual average of the deficit of power supply measured in GWh and from now on called *hydroelectric deficit*.

Finally,as far as flood protection is concerned, the objective is obviously to minimize the damage caused by floods in the city of Como, which occur when the level of the lake exceeds that of the main square (Piazza Cavour). Quantifying this damage is an intricate matter, although many attempts have been made in the past to evaluate the economic value of the damages suffered by the city in floods (the oldest evaluation is that of Cattaneo in 1837). The literature often suggests two indicators of the value of these damages. The first is the "peak of the flood", which considers damages sufffered directly by structures that have come in contact with water. The second is the "duration of the flood", which reveals the indirect damages suffered by the community because of interruptions in traffic, obstructions of public services, and reductions in some production activities. The total damages should, consequently, be evaluated using an appropriate weighted sum of these two indicators. However, a detailed statistical analysis of the available data has shown that these two indicators are strongly correlated, and this has led us to conclude that in the case of lake Como the damages caused on the average by flooding can be expressed simply in function of one of the two, for example, by the mean *number of days of flooding* per year.

The suggestions and costraints reported in the license act approved by the Ministry of Public Works in 1942, shortly before the innauguration of the Olginate dam, attempted to establish a rational compromise among these three objectives. The license act established that the manager is free to decide how much water to supply when the level of the lake lies between two precise limits, $\underline{x}$ and $\bar{x}$, called the lower and the upper limit of the active storage. The interval between the two is called the *operating range*. Beyond these limits the manager must follow a set procedure: at the lower level $\underline{x}$ the flow released from the dam must not exceed the actual inflow into the lake; at levels equal to, or above $\bar{x}$, the regulation dam must be completely opened (free regime). The complex debate that has led to the definition of the license act dates from the early decades of the last century. However, the first systematic hypotheses of operating the lake as a reservoir were formulated only at the end of the century by the industrial users, who had already installed a number of hydroelectric plants some years earlier. Later, the agricultural users added their demands, asserting the priority of irrigation needs. All these requests were amalgamated in 1938 at the moment of the constitution of the Adda Consortium, to which the task of managing the lake has since been entrusted. In the meantime, numerous technical studies were conducted in support of the various requests. They concurred in proposing the damming of the Adda river where it flows out from the lake, in order to store water in periods of elevated inflow - at the beginning of Summer and in the Fall - so that successive periods of deficient flow would be less critical for downstream users. These proposals differed regarding the desirable breadth of the operating range (in particular, suggesting from 1.00 to 1.50 meters at Fortilizio as the upper limit of the active storage) and concerning the suggestions to follow to periodically fill in the lake. Finally, "on the basis of long and most accurate studies", the operating range was set at [-0.50 and 1.20] meters at Fortilizio, recommending a first filling of the lake within the first half of June for use during the Summer, (primarily for agricultural consumption), and a second towards the middle of October for use during the Winter months (prevalently for the hydroelectric plants).

Among the many reactions to the planned regulation, the most significant came from the lakeside communities, who feared worse flooding. In particular, the city of Como, which had always suffered severe damages from floods, suggested in 1927 that the Fall fill be delayed after the tenth of November and then in 1939 that the lower part of Piazza Cavour be raised at the expense of the Adda Consortium. Some studies showed, instead, that high waters in the lake could be attenuated by restructuring some downstream rapids. These studies also expressed the conviction that the operation of the dam actually had little to do with the flooding of Piazza Cavour, in view of the ample clearance between the upper limit of the active storage and the level of the square. This situation has, unfortunately, changed over the years, due to the subsidence of Piazza Cavour. Although there is no precise data in this regard, the phenomenon has been considerable, registering for example some 50 cm between 1967 and 1973. It is obvious that this has aggravated the problem of flooding in Como, rendering it increasingly dependent upon the decisions of the manager, to the point of forcing the latter to gradually modify his historical operating rule, as we shall see.

If at the beginning of every day t the level $x(t)$ of the lake falls in the operating range (that is, $\underline{x} < x(t) < \bar{x}$), the manager must decide how much water to release during the next twenty-four hours. In doing so, he solves as best he can the conflicts among the various objectives that characterize the problem. Naturally, he bases this decision on previous experience, and on all available information regarding the present state of the basin. Thus, the amount of water released $r(t)$ depends first of all upon the day t, which identifies the season and the most probable trend of future inflows (for example, when the season of floods is close, all other conditions being equal, more water will be released than in other seasons). Moreover, the release $r(t)$ is greatly influenced by the level $x(t)$, which represents the quantity of water already available in the lake. Finally, the amount released is also influenced by the conditions of the watershed (snow pack, temperature, humidity, storages of upper reservoirs,...), as well as by the state of the agricultural crops, and any variations forseen in hydroelectric demands. Naturally, these factors can not be modelled formally: some of them were only sporadically present in the historical period, that is, only when the corresponding information was promptly transmitted to the manager. In general, it can be said that the volume released $r(t)$ should be a function of the time t, of the amount of water in the lake $x(t)$, and of a set $y(t)$ of other significant variables, i.e.

$$r(t) = r(t, x(t), y(t))$$

In practice, we may assume that in first approximation the dependence upon $y(t)$ is negligible, or at least weak. We can, therefore, justifiably suppose that in the historical operating period the manager implicitly made his daily decisions on the basis only of his knowledge of t and $x(t)$, that is, according to a function

$$r(t) = r(t, x(t))$$

This function, universally known as *operating rule* , is the basis for a correct statement of the problem of real time operation of a lake, or artificial reservoir, and also serves as a simple operating tool for its management. In fact, once the function $r(t, x(t))$ has been set in a table (for example, for every day and every centimeter of water in the lake), the amount to release during the day can be determined immediately at the beginning of each day: one has only to measure the level $x(t)$ of the lake and find in the table the corresponding value of the flow to release.

In the following section the method for determining the operating rule used implicitly by the manager in the historical period of operation is briefly recalled, while the fourth section describes a new operating rule which, according to the studies made, seems to guarantee a more effective management of the lake.

3 Analysis of the past management

The first step in identifying the operating rule $r(t, x(t))$ used implicitly by the manager during the historical period of operation, was a detailed statistical analysis of the actual inflow of water and the daily distributions from 1946 through 1978. This period of more than three decades can not be considered homogeneous, since over the years the construction of a number of artificial reservoirs in the watershed noticeably modified the inflow regime. The global capacity of the hydroelectric reservoirs in the watershed of the lake rose from 208 million cubic meters in 1946 to 515 million cubic meters in 1978. More precisely, the new reservoirs came on stream between 1957 and 1958. Considering that from 1946 to now there must certainly have been changes in the way the manager operates, and that in the begining he had to learn how to cope with his task, we referred essentially to the years from 1969 through 1978 in identifying the historical operating rule, and we reserved more recent data for testing our results.

The problem of identifying the function $r(t, x(t))$ was first transformed into a parameter estimation problem, selecting beforehand the class of functions to which the function $r(t, x(t))$ must belong. For various reasons the class of piecewise-linear functions was chosen, and the number of segments was set at three to ensure the statistical reliability of the estimates. It was then possible to determine, with the least squares method, the function $r(t, x(t))$ that best interprets the actual management during the historical period from 1969 through 1978. A more detailed analysis showed that the operating rule was gradually modified during that period, due to the manager's growing awareness of the subsidence of Piazza Cavour. The identified operating rule was finally discussed and interpreted, and this further investigation allowed us to conclude that the dependence on the level $x(t)$ of the operating rule used implicitly by the manager is, with reasonable approximation, of the type shown in Figure 2.

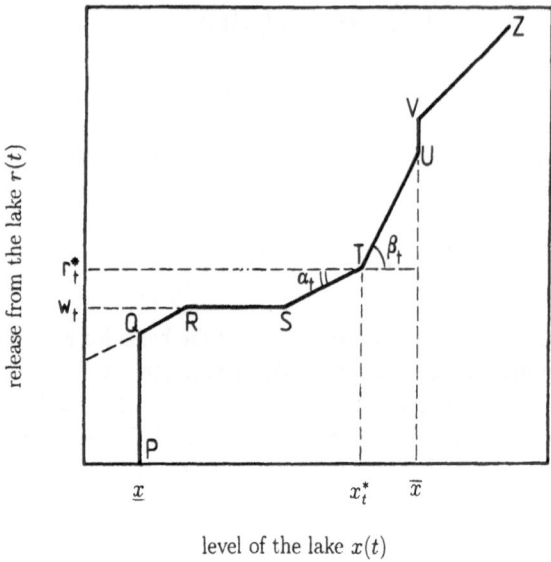

level of the lake $x(t)$

Figure 2. The operating rule of Lake Como

The significance of this rule can be be summarized as follows. In normal conditions of level x_t^* and inflow, the manager follows a preset annual distribution plan r_t^* . Should previous displacements of the inflow from normal values cause the level $x(t)$ of the lake to differ from that of the plan x_t^*, the manager varies the distribution plan by Δr, depending upon the difference between the actual level and the plan level, that is:

$$r(t, x(t)) = r_t * + \Delta r(t, x(t) - x_t *)$$

This variation can, in a first approximation, be expressed in the form (segments ST and TU of Figure 2):

$$\Delta r = \begin{cases} \alpha_t (x(t) - x_t *) & if \quad x(t) < x_t * \\ \beta_t (x(t) - x_t *) & if \quad x(t) > x_t * \end{cases}$$

where the parameters α_t and β_t depend upon the day t and are periodic within the year. These parameters represent the sensitivity of the manager to the agricultural and hydroelectric objectives, and to flooding (for example, the parameter β_t presents sharp peaks in typical flooding periods). For particularly low levels (segment RS in Figure 2), the manager simply releases the *agricultural demand*

$$w_t = \sum_{i=1}^{6} w_i(t)$$

as long as the rules of the license act allow him to do so, and then passes to the free regime (segment QR in Figure 2). In the following the values of the parameters α_t and β_t corresponding to the operating rule applied implicitly by the manager in the historical period are indicated with $\alpha_t{}^*$ and $\beta_t{}^*$, respectively.

4. Real time operation

Once the identification described in the previous chapter had been accomplished, we asked ourselves whether by slightly modifiying the operating rule used by the manager during the historical period, we might be able to improve the management. We decided not to change the characteristic seasonal progression of the parameters, but to proceed by analyzing, instead, the influence of a simple overall decrease, or increase in these parameters. In other words, the operating rules considered acceptable *a priori* (among which we then searched for the efficient ones) were still of the kind shown in Figure 2, but with values of the parameters α_t and β_t proportional to $\alpha_t{}^*$ and $\beta_t{}^*$,characterizing the historical operating rule, that is, $\alpha_t = a\alpha_t{}^*$ and $\beta_t = b\beta_t{}^*$. In this way one has only to determine the optimal values (a°, b°) of the unknown parameters a and b, which assume the significance of *decision variables,* and which can be obtained by formally solving a *three objectives stochastic mathematical programming problem.* An estimate of the value of the three objectives A(agricultural deficit), E (hydroelectric deficit), and P (number of days of flooding) for preset values of the decision variables a and b can easily be obtained by a computer simulation of the system over a sufficiently long period (in this case the data for the period from 1965 to 1979 were used), and the subsequent calculation of the value of the three objectives over that period. It is then possible, applying well known algorithms of multigoal mathematical programming, to determine all the pairs *(a°,b°)* corresponding to *efficient* solutions of the problem formulated above. Of all the acceptable solutions, these are the only ones that enjoy the following fundamental property: *an acceptable real-time operation is said to be efficient when no further improvement can be obtained contemporaneously in all three objectives no matter how the decision variables are altered.* It is also of fundamental importance to see whether the real-time operation corresponding to the decision variables $a = b = 1$ is efficient; if it is not, then it is possible to improve the operation of the lake from all points of view. The graph in Figure 3 shows, in the A,P plane, the efficient solutions with E = const. (191, 192, 195, 198 and 200 GWh).The figure shows how marginal improvements in power production profoundly affect the other aspects of the operation. It also shows that when the annual hydroelectric deficit is relatively high (200 GWh, for example), the flood indicator is almost unaffected by an average agricultural deficit of over 80 million cubic meters. Further reductions in the agricutural deficit imply, instead, extremely high increases in flooding. By contrast, with low levels of hydroelectric deficit, an improvement in the agricultural deficit gradually brings worse flooding. Finally, the figure demonstrates that the operation in the 1969-1978 period was not efficient. In fact, we see immediately that the AB segment corresponding to the "historical" hydroelectric deficit (E = 195 GWh) is characterized by values of A and P below the istorical values corresponding to point C in the figure: for example, flooding could be reduced by 40 per cent, and the agricultural deficit by 48 per cent without altering the hydroelectric deficit (see point P in Figure 3).The same figure also demonstrates explicitly that the three operating objectives can be contemporaneously improved if efficient operating rules, corresponding to the points within the curvilinear triangle ABC, are applied.

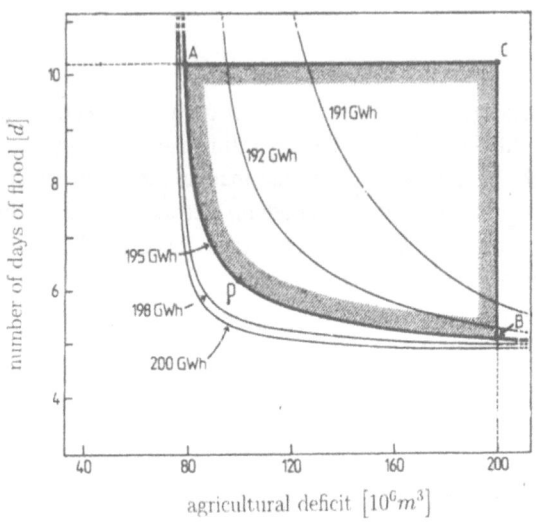

Figure 3. Efficient solutions at constant electric power production

As we have already pointed out, the results obtained can be put to practical use by means of a table in which the volume to release is indicated for every day of the year (line) and for every level of the lake (column). The table corresponding to point P has accordingly been transmitted to the manager. Taking any line of the table at random, we can see that the volume to release follows the trend illustrated in Figure 3, and remains for long segments at a value equal to that of the agricultural demand. If we look, instead, at a column relative to a high level of the lake (for example 110 cm), we see that the release is a bimodal function of time, similar to that of the inflows, with maximum values in May and in the Fall. In these periods the operating rule recommends passing to the free regime even if the lake level falls within the operating range, and this is, indeed, the characteristic which produces a considerable attenuation of the flooding, as shown in Figure 3.

In conclusion, on the basis of this study we can state that improvements can be made from the agricultural and the hydroelectrical point of view, and from that of protecting the city of Como from flooding. The fact that these improvements are achieved by marginal, and not structural, adjustments of the operating rule shows that the decision rules followed by the manager in the past, and in particular, the seasonal fluctuations in his line of action, are, although not optimal, certainly justified. On the basis of these results, we can plausibly anticipate that if the Adda Consortium were to implement the efficient operating rule corresponding to point P of Figure 3, both the flooding and the agricultural deficit would be reduced, without penalizing hydroelectric production.

5. Conclusions

The conclusions drawn in the previous paragraph allow us to look to the future management of Lake Como with optimism. We must also remember that the criteria for efficient management described here do not represent absolute optima: the flooding and agricultural and hydroelectric deficits could be even further contained with the application of more complex operating rules that do not depend on the levels of the lake alone, but include other important information, such as rainfall in the watershed, snow pack, and weather forecast, that more completely describe the state of the watershed. Naturally, this kind of regulation of the lake would require a network that could collect and transmit the data to the manager systematically and in real time. Moreover, an operating rule of this sort could not be summarized in a simple table, because the daily release would depend upon a great number of variables. There would have to be a decision support system that could truly assist the manager in deciding every day the volume of water to release from the regulation dam. For a more detailed development of these matters, the reader is referred to Guariso et al. (1984, 1985).

REFERENCES

Cattaneo, C. (1837), Del Lago di Como, *Eco della Borsa*, n. 17, 18-30 aprile – 7 maggio.

Garofalo, F. – U. Raffa – R. Soncini-Sessa, (1979), Analisi quantitativa della gestione delle acque del lago di Como. Convegno di Idraulica Padana, Magistrato per il Po, Parma, 19-20 ottobre.

Garofalo, F. – U. Raffa – R. Soncini-Sessa, (1980*)*, Identificazione della politica di gestione del lago di Como, XVII Convegno di Idraulica e Costruzioni Idrauliche, Palermo, 27-29 ottobre.

Guariso, G. – S. Rinaldi – R. Soncini-Sessa, (1981), La regolazione ottimale del Lago di Como: analisi a molti obiettivi, "*L'Energia Elettrica*", vol. 58, n.7, settembre.

Guariso, G. – S. Rinaldi – R. Soncini-Sessa, (1985), Decision Support Systems for Water Management: the Lake Como Case Study, "*European Journal of Operational Research*", vol. 21, n. 3, pp. 295-306.

Guariso, G. – S. Rinaldi – R. Soncini-Sessa, (1986), The Management of Lake Como: a Multiobjective Analysis, "*Water Resources Research*", vol. 22, n.2, pp. 109-120.

Guariso, G. – S. Rinaldi –P. Zielinsky, (1984), The Value of Information in Reservoir Management, "*Applied Mathematics and Computation*", vol. 15, n. 2, pp. 165-184.

MECHANICS APPLIED TO THE UNDERGROUND STORAGE OF RADIOACTIVE WASTE MATERIALS

K.S. Chan
Southwest Research Institute, San Antonio, TX, USA

S.R. Bodner
Technion University - I.I.T., Haifa, Israel

D.E. Munson
Sandia National Laboratories, Albuquerque, NM, USA

Abstract: An application of Mechanics to an important environmental problem is an investigation on the creep properties of rock salt which was undertaken in relation to the planned encapsulation of transuranic nuclear waste in caverns excavated in bedded salt formations (the WIPP program in the USA). Those caverns are intended to serve as permanent repositories for radioactive waste over an extensive period so that complete isolation is required of the facility including the shafts that are initially connected to the outside. In conjunction with an extensive experimental program, the analytical and numerical studies on creep of rock salt were concerned with the following subjects: creep based on dislocation mechanisms; damage induced creep leading to volumetric changes, pressure dependence, and creep rupture; healing of damage; failure and fracture mechanisms; and structural integrity of underground storage rooms

Introduction

Permanent storage of high level radioactive waste has become a major environmental problem throughout the world and has motivated extensive investigations. Deep embedment of waste materials in appropriate geological formations in mountains and in the earth is a direction undertaken in many countries. Mechanics has a large role in the investigations on the long time security and reliability of the means of storage and this paper is a brief review of an analytical and experimental investigation related to a project in the U.S.

The methods for long time storage that are being examined in Europe generally involve initial encapsulation of the waste materials in special impermeable and radiation resistant containers such as POLLUX casks. Methods for permanent deposition include,

(a) placement of the containers longitudinally on the floor of excavated horizontal drifts of rock salt, about 800-1000 m below the surface, and complete coverage of the containers to the ceiling by dry crushed salt, and

(b) placement of the containers vertically in cylindrical cavities bored in the floor of tunnels in hard rock formations, followed by surrounding the containers by highly compacted bentonite, a high density concrete, and backfilling of the tunnels and shafts.

[1] Work performed at Southwest Research Institute, San Antonio, Texas, USA, supported by the U.S. Department of Energy (DOE).

Repository projects in the U.S. are the Yucca Mountain, Nevada, program which also involves special containers to encapsulate the waste materials and their storage in tunnels bored in a mountain of hard rock, and the WIPP program in southeastern New Mexico for permanent storage in natural bedded formations of rock salt deep below the ground surface. For the WIPP (Waste Isolation Pilot Plant) program, a large number of caverns have been excavated 650 m below the surface which are 5.5×5.5 m in cross section and about 93 m in length. Radioactive waste in standard containers will fill the cavern and complete encapsulation and isolation would be a consequence of creep deformations of the surrounding rock salt after a number of years. Application of this proposal requires good understanding of the mechanical behavior of rock salt since the caverns should be operative and available for about 100 years prior to their closure by creep deformations.

It was found that damage in the form of wing cracks is an important factor in the creep characteristics of rock salt. Development of damage directly contributes to the creep strains and also leads to volumetric changes and to pressure dependence of the deformations. Rock salt is a semi-brittle material and damage incurred during creep straining could result in material failure. Unlike the behavior of metals, healing of damage in rock salt could occur under appropriate loading and temperature conditions.

A number of papers and a review article have been published on the analytical and experimental investigations of the mechanical properties of rock salt related to the WIPP program and only some of the main points are described here. The analytical work was performed at Southwest Research Institute in San Antonio, Texas, under contract from the Sandia National Laboratories which is the scientific support contractor to the U.S. Department of Energy on the project. Aside from that investigation, other projects, especially in Germany and Canada, have generated a large literature on the creep deformation and possible failure of rock salt, e.g. Aubertin et al. (1993a,b), Cristescu and Hunsche (1992), Cristescu (1993), Cristescu and Gioda (1994), Pudewills et al. (1995), Pudewills (1998).

Formulation of Constitutive Equations

In the analytical formulation, the total strain rate $\dot{\varepsilon}_{ij}^t$ for a solid deformed under isothermal conditions is given as the sum of the elastic strain rate $\dot{\varepsilon}_{ij}^e$ and the inelastic strain rate $\dot{\varepsilon}_{ij}^I$ which are both generally non-zero, i.e. a yield criterion is not specified. The elastic strain rate components are given by the time derivative of Hooke's Law. The starting point for the development of an expression for the inelastic strain rate was a set of isotropic constitutive equations for creep of undamaged rock salt developed by Munson and Dawson (1984) based on dislocation mechanisms. As such, the equations do not indicate pressure dependence and anisotropy of inelastic straining or volume changes.

Those effects and others required consideration of "damage" in the material equations which consisted of introducing a scalar damage parameter ω as a "softening" internal variable, in the sense of Kachanov (1958), with an evolution equation for its development. Generation of damage would also lead directly to additional inelastic straining, pressure dependent dilatancy and possible failure. These effects and the possibility of damage recovery are represented analytically by a generalized flow law for inelastic straining, [Chan et al. (1992, 1998, 1999)]:

$$\dot{\varepsilon}_{ij}^l = \frac{\partial \sigma_{eq}^c}{\partial \sigma_{ij}} \dot{\varepsilon}_{eq}^c + \frac{\partial \sigma_{eq}^{\omega_s}}{\partial \sigma_{ij}} \dot{\varepsilon}_{eq}^{\omega_s} + \frac{\partial \sigma_{eq}^{\omega_t}}{\partial \sigma_{ij}} \dot{\varepsilon}_{eq}^{\omega_t} + \frac{\partial \sigma_{eq}^h}{\partial \sigma_{ij}} \dot{\varepsilon}_{eq}^h \qquad (1)$$

where $\sigma_{eq}^c, \sigma_{eq}^{\omega_s}, \sigma_{eq}^{\omega_t}, \dot{\varepsilon}_{eq}^c, \dot{\varepsilon}_{eq}^{\omega_s}$, and $\dot{\varepsilon}_{eq}^{\omega_t}$, are power-conjugate equivalent stress measures and equivalent inelastic strain rates. The power-conjugate stress measures are functions of the current principal stress components and operate as flow potentials. Their derivatives with respect to the stress components provide the direction of inelastic straining for each respective mechanism. The corresponding equivalent inelastic strain measures supply the magnitudes of straining and are functions of the associated stress measures, temperature, and internal state variables. If the stress measures used in those functional relationships differ from those that provide the direction of inelastic straining, then the formulation is referred to as "non-associated" which can be relevant for an accurate description of rock salt behavior, Chan et al. (1994). It was found necessary to distinguish between the stress measures of shear damage, $\sigma_{eq}^{\omega_s}$, and that of damage due to the maximum tensile stress, $\sigma_{eq}^{\omega_t}$, since the nature of the stress leads to different geometries of microcracking and corresponding straining. The kinetics of damage development was also different for shear and tensile loading conditions. Consequently, the expressions for the damage induced equivalent inelastic strain rates differ for shear, $\dot{\varepsilon}_{eq}^{\omega_s}$, and tension, $\dot{\varepsilon}_{eq}^{\omega_t}$. The equivalent stress measure for dislocation creep σ_{eq}^c corresponds to the Munson-Dawson (1984) equations with the inclusion of the damage parameter as a softening effect. The parameters represented by σ_{eq}^h and $\dot{\varepsilon}_{eq}^h$ are the conjugate equivalent stress and strain rate measures for damage healing.

Equation (1) indicates a directional dependence of the inelastic strain rate on the stress state so that inelastic straining is not isotropic even though a scalar damage parameter is utilized. Formulation of the appropriate expressions for the stress measures due to damage development should be governed by the physics of the processes by which straining develops. In the case of damage due to shear, wing cracks occur at the tips of the microcracks induced by slip and it is these wing cracks that promote the additional strains, e.g. Fig. 1. The expression for $\sigma_{eq}^{\omega_s}$ consists of a term for damage generation and related creep, namely, the maximum shear stress which would be the maximum difference of principal stresses, $|\sigma_1 - \sigma_3|$, and a term for suppression of damage development and ensuing creep due to the confining pressure P. A tensile principal stress would increase damage growth with consequent straining so that $\sigma_{eq}^{\omega_t}$ consists only of the tensile stress component multiplied by a constant. The formulation by means of Eq. (1) therefore includes non-isotropy and pressure dependence of inelastic straining as well as volumetric expansion under non- hydrostatic triaxial compressive loading.

The material constants in the analytical model were determined by fitting model calculations to experimental creep curves of rock salt from the Waste Isolation Pilot Plant (WIPP) site. The material constants for clean and argillaceous (clay bearing) WIPP salt are presented in the various references and have been used for the calculations shown here where they are referred to as the MDCF (Multi-mechanism Deformation Coupled Fracture) Model.

Coupling of creep and damage in the analytical model allowed calculation of the entire creep curve including tertiary creep. Calculated creep curves for WIPP clean salt tested at a stress difference of 25 MPa under a confining pressure of 1 or 15 MPa are compared against the experimental data in Fig. 2, from Chan et al. (1994). The agreement between model calculation and experimental data is considered good because it is within the variability factor of two usually observed in the experimental creep curves.

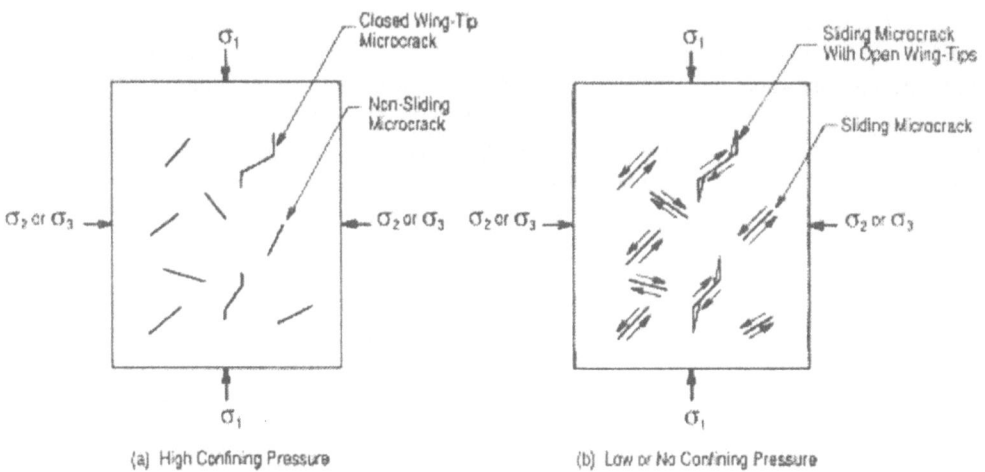

Figure 1. Damage mechanisms envisioned during creep deformation of rock salt under triaxial compression: (a) closure of microcracks by a high confining pressure; and (b) generation and opening of microcracks at low or no confining pressure. Sliding or shearing of microcracks leads to the deviatoric component, while opening of wing-tip microcracks lead to the dilatational component of the damage-induced inelastic strain rate, $\dot{\varepsilon}_{eq}^{\omega}$.

Part of the formulation is the evolution equation for damage development. This follows the general form used by most investigators in which the rate of damage generation is a function of damage multiplied by a function of stress which contains the equivalent stress measures for shear and tensile damage used in the kinetic equations for the equivalent inelastic strain rates. Details are given in Chan et al. (1994, 1997a,b, 1998, 1999).

It is noted that the procedure described above utilizes the coordinates for the current principal stress state and not coordinates fixed in the material. This provides for coordinate invariance as well as for numerical convenience. Calculated creep deformations in the current principal stress directions could be translated incrementally to bodily fixed coordinates for non-proportional stress histories. Within this context, treating damage as a second or fourth order tensor appears to be complicated and not necessarily more accurate for the calculation of inelastic strains and deformations.

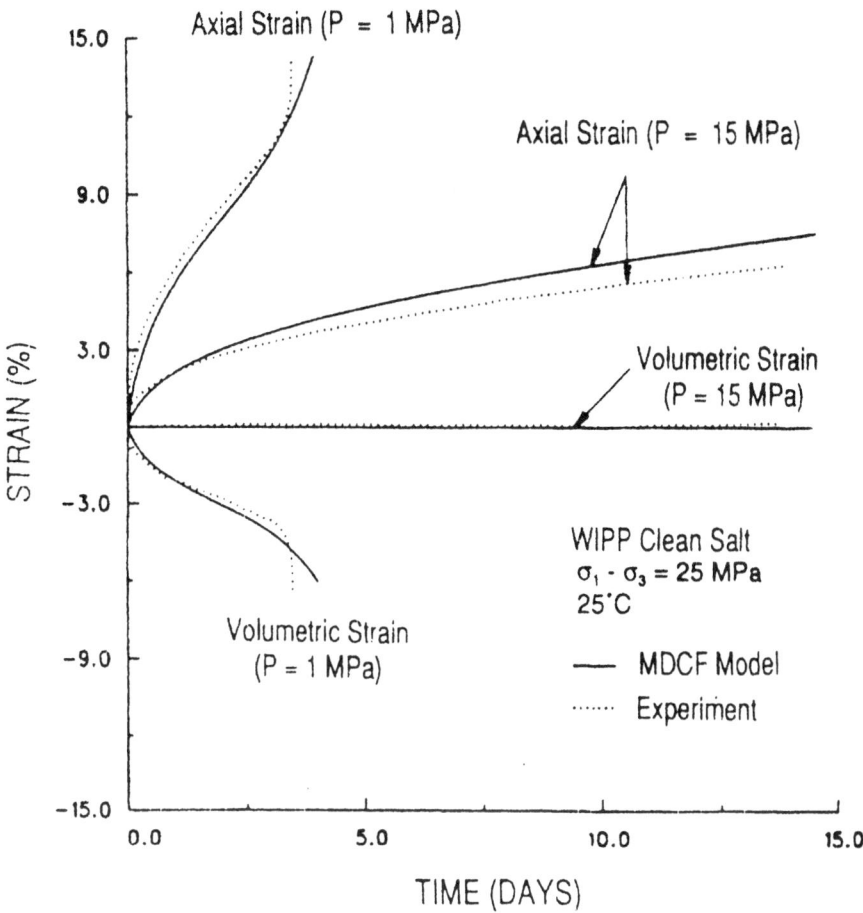

Figure 2. Experimental creep curves of WIPP salt tested at $\sigma_1 - \sigma_3 = 25$ MPa under a confining pressure, P, of 1 or 15 MPa with comparison to model calculations. Experimental data are from Senseny (1986) and Fossum et al. (1993) [from Chan et al. (1994)].

Recovery and healing of damage developed in rock salt would be expected to occur under hydrostatic pressure and moderate to high temperatures. These conditions would lead to the closure of wing cracks and voids and at least partial merging of crack surfaces by a sintering like process. Since the microcracks induced by the prior loading would have directional characteristics, the straining associated with damage recovery and healing would also not be isotropic. The last term in eq. (1) is intended to account for recovery and healing strain rates,

$$\dot{\varepsilon}_{ij}^{h} = \frac{\partial \sigma_{eq}^{h}}{\partial \sigma_{ij}} \dot{\varepsilon}_{eq}^{h} \tag{2}$$

where σ_{eq}^{h} is the equivalent stress measure associated with the generation of healing strains and $\dot{\varepsilon}_{eq}^{h}$ is the equivalent healing strain rate measure.

A simple form that accounts for the anisotropy of straining due to healing is,

$$\sigma_{eq}^{h} = \frac{1}{3}(I_1 - B\sigma_1) \geq 0 \tag{3}$$

where I_1 is the first stress invariant, B is a constant, and σ_1 is the maximum principal stress with compression being positive. For hydrostatic compression, the first three terms in eq. (1) become zero, since creep and formation of microcracks would not occur, while the last term, from eq. (3), would be non-zero. It would also be non-zero for compressive stress states other than hydrostatic. In the case of a cylindrical specimen under hydrostatic compression, eqs. (2) and (3) lead to

$$(\dot{\varepsilon}_{11} / \dot{\varepsilon}_{22}) = 1 - B \tag{4}$$

for the ratio of axial to lateral strain rates of healing.

Tests of the healing of pre-damaged cylindrical specimens indicate that the actual strain rate ratio is bilinear with time so that two different mechanisms, each with a different value of B, appear to be operating during the healing process. Plots of changes of volumetric strain as a function of time also indicate that one mechanism acts at short times and another at longer times. For each of these mechanisms, the equivalent strain rate measure $\dot{\varepsilon}_{eq}^{h}$ can be represented by a first order kinetic equation and a characteristic time constant. Details of the analysis are given in Chan et al. (1995a, 1998).

From the physical viewpoint, the initial short time mechanism for which B>1 at all test temperatures could be related to the closure of open wing cracks aligned in the axial direction with consequent axial extension. For the second longer time mechanism, B=1 for all the tests indicating, from eq. (4), no axial straining during that stage of the healing process. Volume change still occurs for the second mechanism which could correspond to the merging of the surfaces of axially aligned cracks. A plot of measured strain changes during the healing process and corresponding analytical predictions are shown in Fig. 3, from Chan et al. (1995a). It is noted in this example that lateral and volumetric dimensions reduce during the healing process while the axial dimension initially increases and then remains constant. The non-isotropy of the actual healing process seems to be reasonably well represented by the analytical model.

In addition to generating straining, healing of damage would reduce the amplitude of the damage variable. A damage healing term therefore needs to be added to the evolution equation for damage; such a term was proposed by Brodsky and Munson (1994).

Tests were also performed on the changes in the ultrasonic wave speeds during the healing process which were then compared to the original velocity of the undamaged material. Velocity measurements were taken both parallel and perpendicular to the axis of cylindrical specimens. At 70°C, the changes in the reduced velocities with time of healing showed the same characteristics in both directions, Fig. 4 in Chan et al. (1995a). At 20°C, however, the tests showed a definite difference indicating the increased effect of damage in reducing the wave velocity in the perpendicular direction, i.e. normal to the crack openings, Fig. 4. A simplified analytical model for damage reduced wave velocity was developed based on considering damage as a scalar. Whereas the analysis provided reasonable results for wave speed recovery parallel to the axis, Fig. 4, it could not account for the anisotropic effect of damage on the wave velocity in the perpendicular direction.

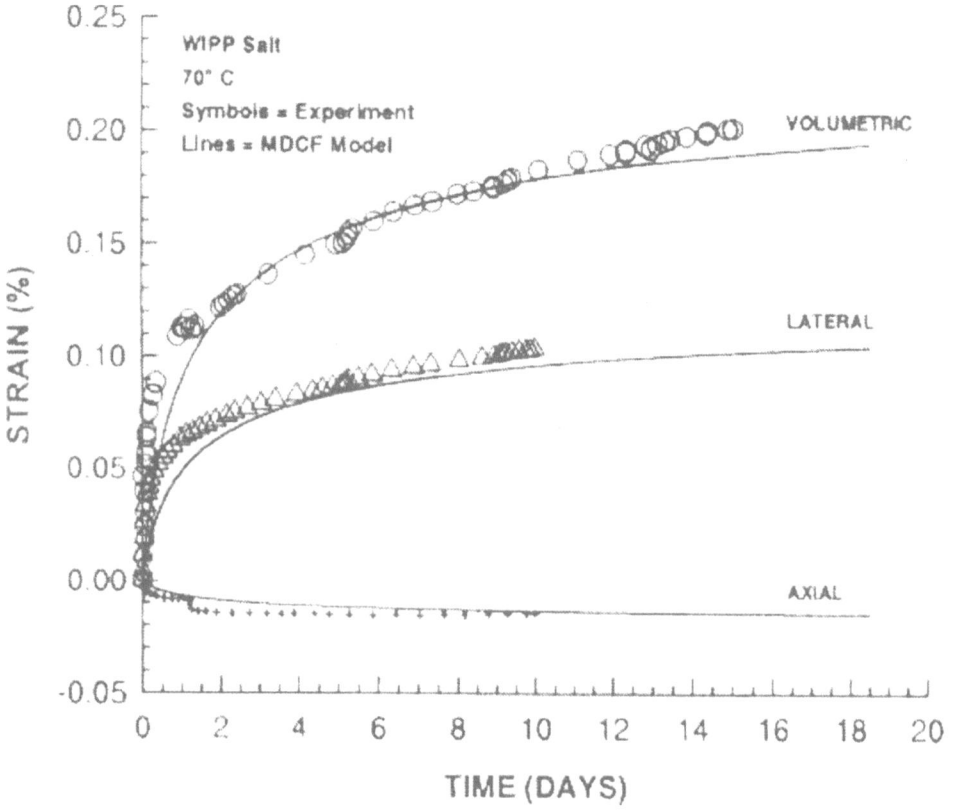

Figure 3. Comparison of calculated and measured volumetric, lateral, and axial strains recovered during damage healing of WIPP salt under a hydrostatic pressure of 15 MPa at 70°C.

A more sophisticated analytical treatment of damage as a second order tensor would be straightforward for the case of fixed principal stress and damage axes that coincide with the geometrical axis of cylinders which are subject to hydrostatic pressure. On such a basis, better agreement in both directions could have been achieved for the case shown in Fig. 4. However, damage effects are particularly important in structures subjected to non-proportional loading histories for which the treatment of damage as a higher order tensor would be appreciably more complicated.

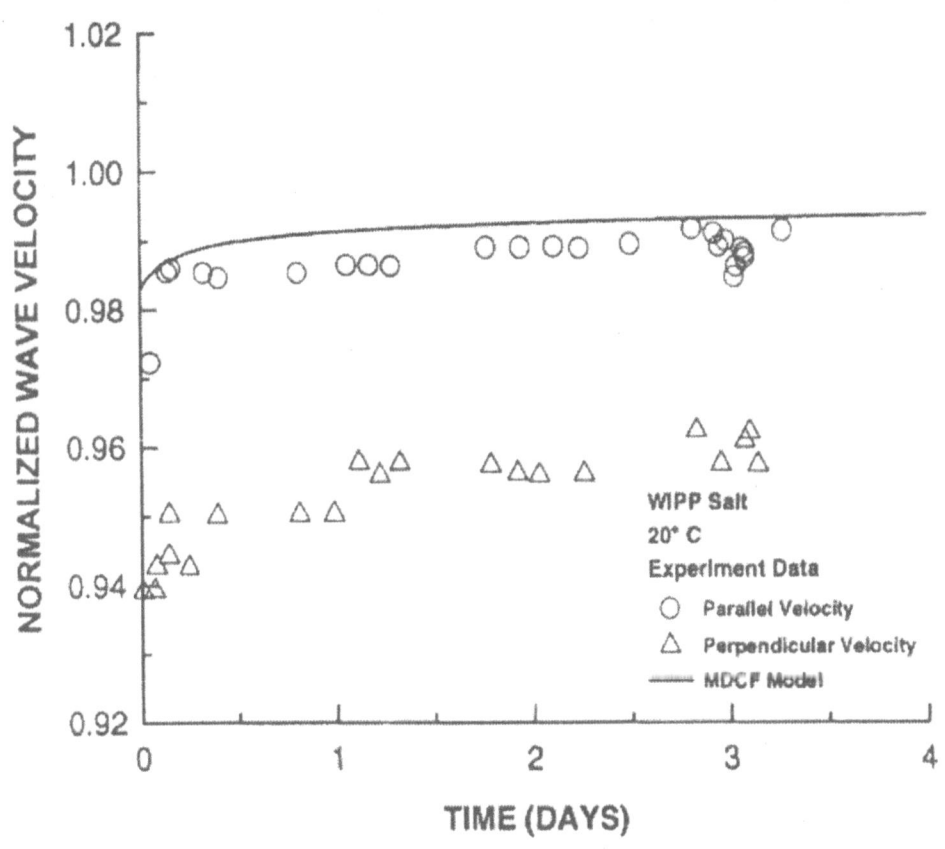

Figure 4. Comparison of calculated and measured wave velocities during healing of salt at 20°C.

Creep Failure and Cleavage

Similar to the early work of Kachanov and of Rabotnov, the inclusion of damage effects in the governing equations for creep of rock salt can lead to tertiary creep and failure. For that to be possible at some unspecified time, the volumetric expansion and damage development rates should be positive which corresponds to a condition on the stress measures. In the particular case of triaxial compression under which most tests have been performed, this condition can be expressed in terms of $\sqrt{J_2}$ and I_1, where J_2 is the second invariant of deviatoric stress and I_1 is the first stress invariant. That relation, as shown in Fig. 5, from Chan et al. (1997a), agrees well with test data.

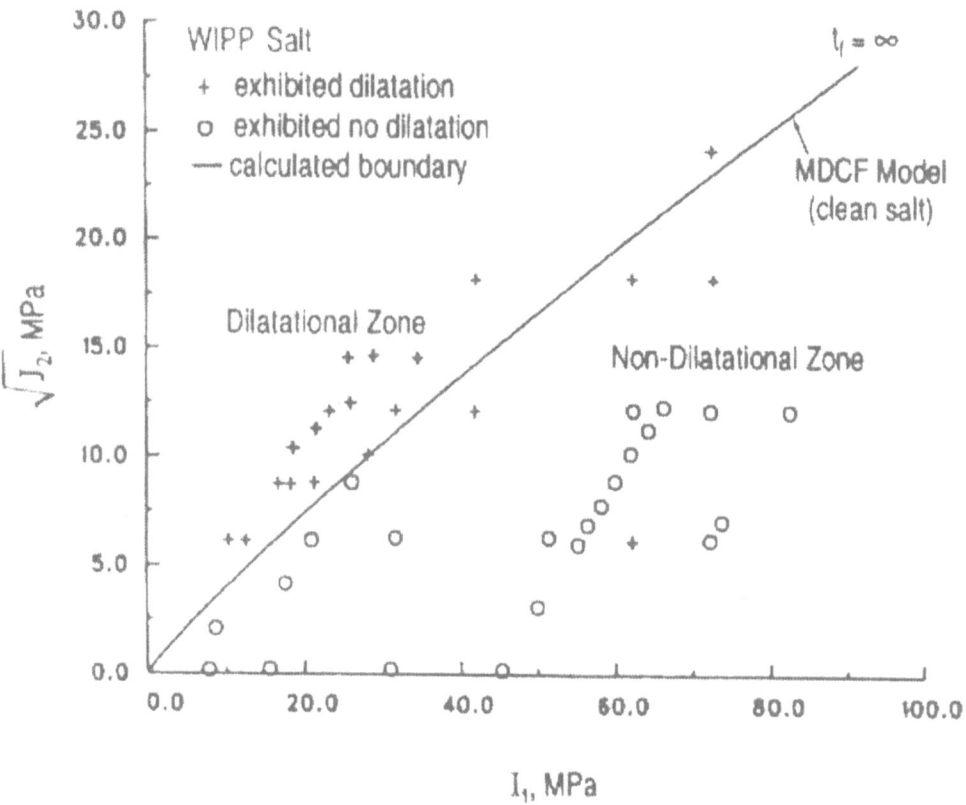

Figure 5. Calculated dilatational zone boundary compared to experimental data compiled by Van Sambeek et al. (1993, and Brodsky and Munson (1994). [From Chan et al. (1997a)].

Elapsed times for the onset of tertiary creep and for failure are important parameters for engineering purposes. Comparisons of numerical simulations based on the proposed model with test results indicate that the onset of tertiary creep generally correlated with an accumulated damage value of 0.015 and that failure occurred at a damage value of about 0.15, Fig. 6. Of various possible criteria for determining creep failure times, the one using a critical damage value of 0.15 proved to be the most consistent with test results, Chan et al. (1997a). The time to rupture increased with increasing confining pressures. At a confining pressure greater than 5.8 MPa, damage development was almost totally suppressed and the time to rupture became extremely large ($> 10^8$ days), as shown in Fig. 6.

Another possible failure mechanism of rock salt is by cleavage which has to be examined by the methods of fracture mechanics. This has been the subject of an investigation which has been reported by Chan et al. (1996b). An investigation of cleavage due to indirect tension was reported by Chan et al. (1997c). One result of the fracture mechanics analysis is that unstable wing crack extension could occur only in the presence of a tensile stress acting normal to the wing cracks. Combinations of stress components that govern the various possible failure modes are indicated on fracture mechanism maps in Chan et al. (1996b). An example is shown in Fig. 7.

An interesting subject is the relationship of elastic moduli reductions due to damage development to the onset of failure. There have been extensive investigations on the reductions of elastic moduli due to the presence of microcracks. These studies have generally been based on particular forms of cracks and their geometrical distribution in the material, i.e. whether in regular arrays or random, and whether or not the cracks are interacting. The concept of a crack density function has been found useful in these analyses. An obvious result is that a preferred orientation of cracks leads to non-isotropy and anisotropic reduced elastic moduli. A thorough review of investigations on the subject up to 1992 was prepared by Kachanov (1992).

The matter of whether a correlation exists between the tendency to fracture of a brittle solid and the change of effective reduced moduli influenced by cracking was raised by Kachanov (1992). Such a correlation would be highly desirable if progression toward fracture could be monitored by the changes of elastic moduli. Kachanov (1992) notes that failure by cleavage is a local effect controlled by the local stress state, the details of the crack geometry, and the fracture toughness of the material. As such, the tendency to cleavage failure would not be expected to necessarily correlate with reductions in elastic moduli which are an overall volumetric average effect. In the case of failure by creep rupture, however, which was not discussed by Kachanov (1992), the correlation could exist since both can be described, at least as an approximation, as dependent on the same scalar damage variable.

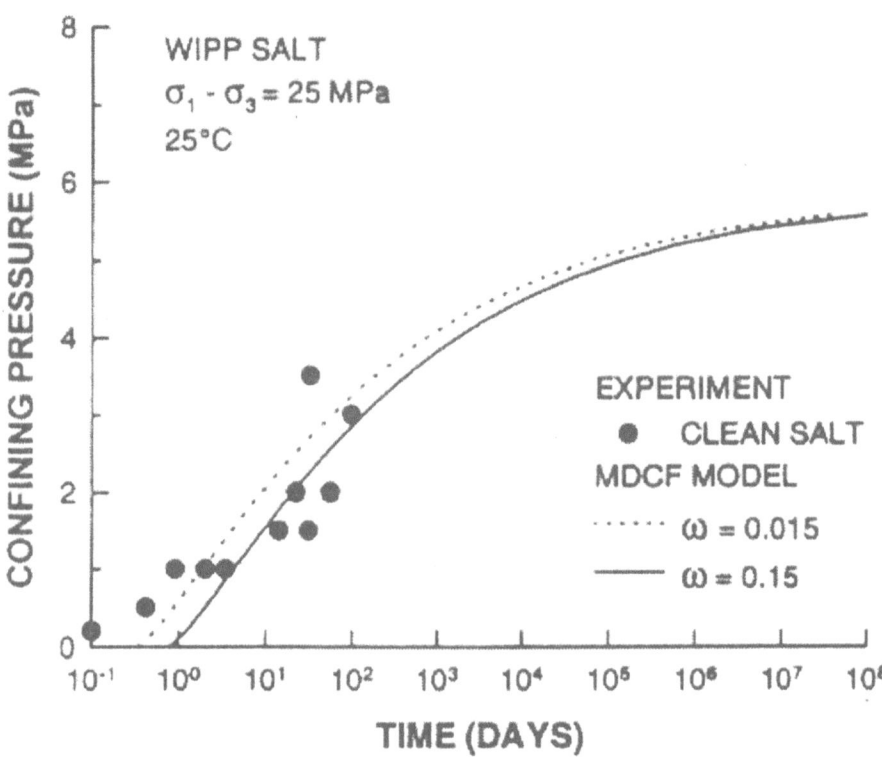

Figure 6. Comparison of calculated and measured creep rupture times for WIPP salt as a function of confining pressure. Creep rupture was take to occur at $\omega = 0.15$ in the model calculations. [From Chan et al. (1997a)].

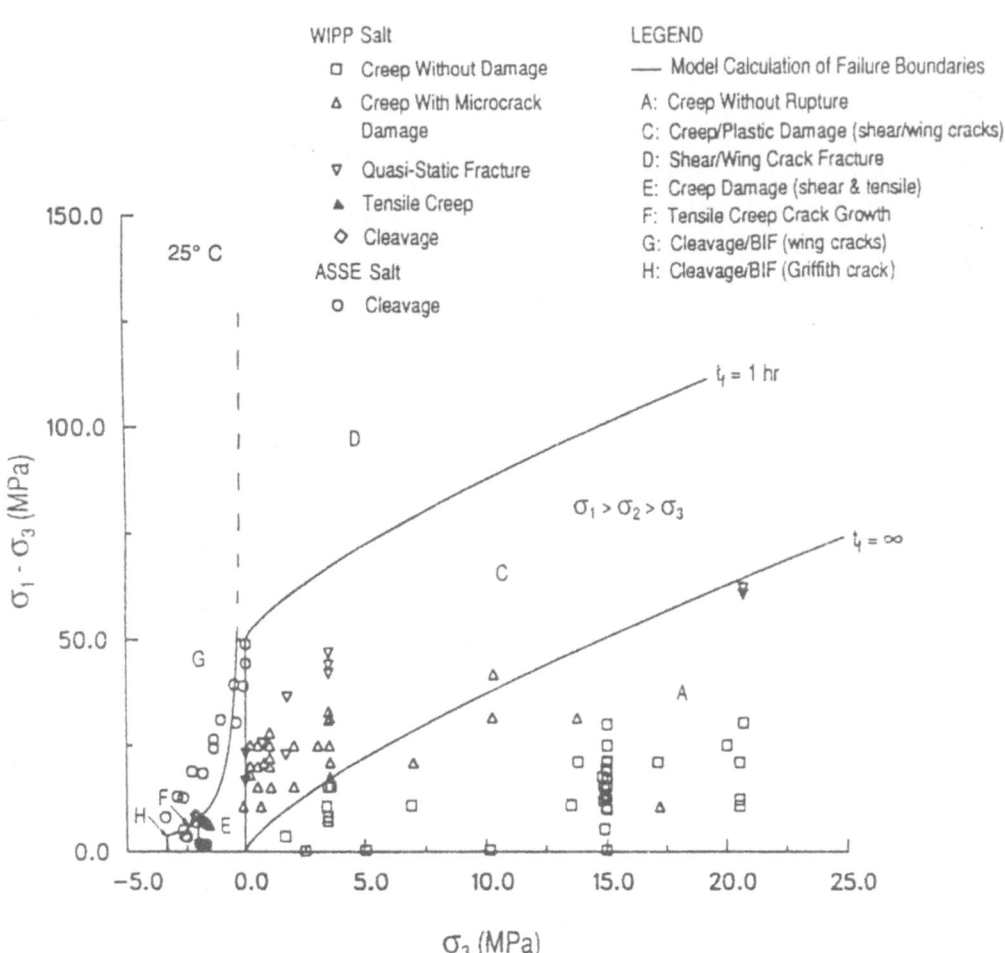

Figure 7. Computed fracture mechanism map compared against experimental data of WIPP salt [Brodsky (1995); Van Sambeek et al. (1993); Senseny (19086); Fossum et al. (1993); Wawersik and Hannum (1979)] and ASSE salt [Hunsche (1993)]. [From Chan et al. (1996b).]

Life Prediction for a Salt Structure

The WIPP is a research and development facility whose purpose is to demonstrate the safe management, storage, and eventual disposal of nuclear waste. The underground portion of the WIPP facility is approximately 650 meters below the ground surface in massive natural deposits of the Salado Formation in southeastern New Mexico. Pure and argillaceous halite materials are present at the WIPP, together with clay seams, polyhalite beds, and anhydrite beds.

At an early construction phase of the WIPP project, a four-room complex was excavated for test purposes. The rooms were 4.0 m in height, 10.0 m in width, 91.4 m in length, and were separated by pillars 30.5 m wide. Each of the rooms was instrumented with several closure stations and extensometers. One room was supported while the other three rooms were intentionally unsupported so that the long-term performance of these unoccupied, barricaded rooms could be studied. Closure of these test rooms by creep deformations eventually resulted in the loss of structural integrity. Figure 8 illustrates the typical fracture pattern that developed in the unsupported experimental rooms.

The constitutive model, referred to as the MDCF model, was incorporated into the finite element code SPECTROM-32, Callahan et al. (1990), which was utilized to analyze the closure response of the test rooms using material constants for WIPP clean and argillaceous salt. This exercise is reported by Chan et al. (1995b, 1999). The four-room complex was modeled as a single, one-half room with vertical symmetry planes through the center of the room and the center of the pillar. A lithostatic initial stress state that varied linearly with depth was assumed and with its value based on the average material density. The calculated initial hydrostatic pressure at the repository horizon was 14.8 MPa. To represent the overburden, a 13.57 MPa normal traction boundary was applied to the upper boundary of the structure. Other boundary conditions were (1) zero horizontal displacements on the symmetry plane and on the right boundary, and (2) zero vertical displacements along the bottom boundary. The reference time for the calculations was set to zero at the time of the complete excavation of the room; the response of the room was calculated for a 10 year closure period at a constant temperature of 300K. The average clay content for the argillaceous salt was 2.9%. For argillaceous salt, the equations are the same as those described here and the material constants are given by Chan et al. (1996a, 1999). In the calculations presented, only the creep and shear damage terms were operative while the tensile damage and healing terms were not used.

The finite element calculations provided results of the room closure rates in the vertical and horizontal directions, the distributions of local stresses, volumetric strains, and values of the damage variable, ω. Only results of the damage variable are presented here to illustrate the use of the continuum damage mechanics approach for predicting the failure response and time-to-failure of the test rooms.

In the formulation, an extremely small initial value of ω, ω_0, was assumed as 0.0001 everywhere in the salt structure at time zero. After excavation, damage developed slowly beginning at the corners of the room where stress concentrated. As time elapsed, the damage zone extended outward to form arch-shaped bands that spanned between room corners. The formation and spread of the damage zones with time is illustrated in Fig. 9 which shows contours of constant value of the damage variable, ω, for the test room at various times after excavation. Contours of higher ω values indicate regions with a higher level of creep damage. After 0.5 year of creep, arch-shaped damage zones with an ω value in the range of 0.0001 - 0.00015 formed between room corners, as shown in Fig. 9. Additionally, high levels of damage with ω values of 0.015 and 0.15 developed at the corners of the room. Damage contours of an ω value 0.015 indicate tertiary creep regions. For ω>0.15, the contours depict ruptured regions, for example, as shown in the region below the floor. With increasing elapsed times, both the tertiary creep zone (ω>0.015) and the rupture zone (ω>0.15) expanded outward from the corners. After a 10 year period, an arch shaped tertiary creep zone was calculated to be fully developed in the roof and pillars and under the floor of the room. It is noted that the ubiquitous damage model produces a greater spread in the damage zone than observed because localization was not included in the calculations.

A comparison of the results shown in Figs. 8 and 9 indicates that the locations of the calculated damage zones and fracture pattern are in good agreement with the *in-situ* field observations. A dish-shaped fracture extended from the floor near the corners of the room through the salt and into the anhydrite below the room. The fracture through the argillaceous halite intersected the floor of the room at approximately the same site where the maximum damage was predicted by the finite-element method (FEM) calculations.

In terms of failure time, the calculated time of fracture formation in the floor was about 10 years after excavation, but was somewhat longer in the roof. In comparison, Room 1 showed signs of instability six years after excavation, and the roof collapsed at approximately eight years after excavation. The rock fall resembled a dish-shaped slab approximately 10 m wide by 2 m high and 45 m length. The roof of Room 2 fell 11 years after excavating. The roof of the other unsupported room had not failed after 11 years when the room was closed in 1995, but its failure was imminent because of increasing room closure rates. Overall, the calculated failure times for the test rooms are reasonably good estimates.

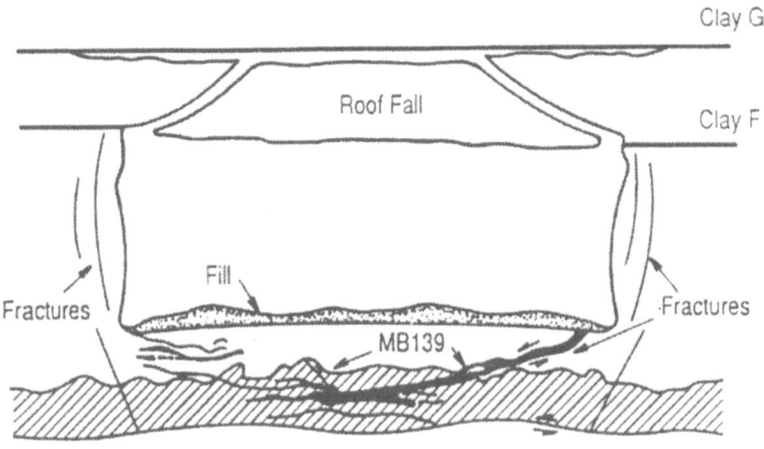

Figure 8. Composite schematic of observed fracture pattern near unsupported WIPP experimental rooms [Chan et al. (1995b)].

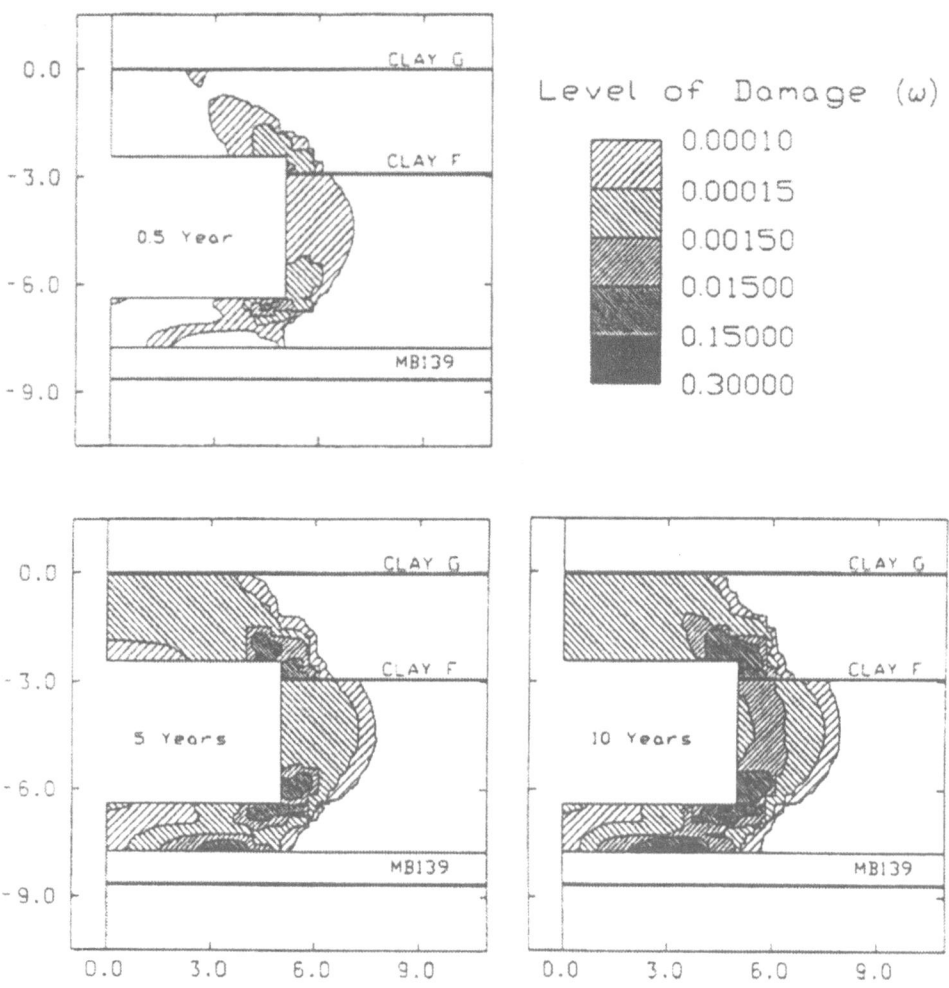

Figure 9. Damage level contours near the test room at 0.5, 5, and 10 years
[Chan et al. (1995b)].

References

Aubertin, M., Sgaoula, J. and Gill, D.E. (1993a). A damage model for rocksalt: application to tertiary creep. In H. Kakihan, H.R. Hardy., Jr., T. Hoshi and K. Toyodura, eds., *Seventh Symposium on Salt.* New York: Elsevier Science Publishers. 117-125.

Aubertin, M., Gill, D.E. and Ladanyi, B. (1993b). Modeling the transient inelastic flow of rocksalt, In H. Kakihan, H.R. Hardy., Jr., T. Hoshi and K. Toyodura, eds., *Seventh Symposium on Salt.* New York: Elsevier Science Publishers. 93-104.

Brodsky, N.S. and Munson, D.E. (1994). Thermomechanical damage recovery parameters for rock salt from the Waste Isolation Pilot Plant. In Nelson, P.P. and Lauback, eds., *Proceedings of the First North American Rock Mechanics Symposium.* Brookfield, VT: Balkema. 731-738.

Callahan, G.D., Fossum, A.F. and Svalstad, D.K., (1990). Documentation of SPECTROM-32: A finite thermomechanical stress analysis program, 1 and 2. RE/SPEC Inc., Rapid City, SD, RSI-0269.

Chan, K.S., Bodner, S.R., Fossum, A.F. and Munson, D.E. (1992). A constitutive model for inelastic flow and damage evolution in solids under triaxial compression. *Mechanics of Materials.* 14:1-14.

Chan, K.S., Brodsky, N.S., Fossum, A.F., Bodner, S.R. and Munson, D.E. (1994). Damage-induced nonassociated inelastic flow in rock salt. *International Journal of Plasticity.* 10:623-624.

Chan, K.S., Bodner, S.R., Fossum, A.F. and Munson, D.E. (1995a). Constitutive representation of damage healing in WIPP salt. In J.J.K. Daeman and R.A. Schultz. eds., *Proceedings of the 35th U.S. Symposium on Rock Mechanics.* Rotterdam, Netherlands: Balkema Publishers. 485-490.

Chan, K.S., DeVries, K.L., Bodner, S.R., Fossum, A.F. and Munson, D.E. (1995b). A damage mechanics approach to life prediction for a salt structure. *Computational Mechanics '95.* 1:1140-1145.

Chan, K.S., Munson, D.E., Fossum, A.F. and Bodner, S.R. (1996a). Inelastic flow behaviors of argillaceous salt. *International Journal of Damage Mechanics.* 5:292-314.

Chan, K.S., Munson, D.E., Bodner, S.R. and Fossum, A.F. (1996b). Cleavage and creep fracture of rock salt. *Acta Metallurgica et Materialia.* 44:3553-3565.

Chan, K.S., Bodner, S.R. Fossum, A.F. and Munson. (1997a). A damage mechanics treatment of creep failure in rock salt. *International Journal of Damage Mechanics.* 6:121-152.

Chan, K.S., Bodner, S.R. and Munson, D.E. (1997b). Treatment of anisotropic damage development within a scalar damage formulation. *Computational Mechanics.* 19:522-526.

Chan, K.S., Brodsky, N.S. Fossum, A.F., Munson, D.E. and Bodner, S.R. (1997c). Creep-induced cleavage fracture in WIPP salt under indirect tension. *ASME Transactions, Journal of Engineering Materials and Technology.* 119:393-400.

Chan, K.S., Munson, D.E. and Bodner, S.R. (1998). Recovery and healing of damage in WIPP salt. *International Journal of Damage Mechanics.* 7:143-166.

Chan, K.S. Munson, D.:E. and Bodner, S.R. (1999). Creep deformation and fracture in rock salt. In Aliabadi, M.H., ed., *Fracture of Rock.* Ashurst, Southampton, UK.: Computation Mechanics Publications.

Cristescu, N. and Hunsche, U. (1992). Determination of nonassociated constitutive equations of rock salt from experiments. In Besdo, D. and Stein, E., eds., *Finite Inelastic Deformation Theory and Application.* Berlin:Springer. 511-523.

Cristescu, N.D. (1993). A general constitutive equation for transient and stationary creep of rock salt. *International Journal of Rock Mechanics, Mineral Science and Geomechanical Abstract.* 30:125-140.

Cristescu, N.D. and Gioda, G. (1994). *Visco-Plastic Behaviour of Geomaterials.* New York:Springer-Wien (CISM Course).

Fossum, A.F., Brodsky, N.S., Chan, K.S. and Munson, D.E. (1993). Experimental evaluation of a constitutive model for inelastic flow and damage evolution in solids subjected to triaxial compression. *International Journal of Rock Mechanics and Mining Sciences and Geomechanics Abstracts.* 30:1341-1344.

Hunsche, U.E. (1993). Failure behavior of rock salt around underground cavities. In H. Kakihan, H.R. Hardy, Jr., K. Koyodura and T. Hoshi, eds., *Proceedings of the Seventh International Symposium on Salt*. New York: Elsevier Science Publishers. 59-65.

Kachanov, L.M. (1958). On creep rupture time. *Otdelenie Technicheskikh Nauk, Izvestiya Akademii Nauk SSSR*. 8:26-31.

Kachanov, M. (1992). Effective elastic properties of crack solids: Critical review of some basic concepts. *Applied Mechanics Reviews*. 45:304-335.

Munson, D.E. and Dawson, P.R. (1984). Salt constitutive modeling using mechanism maps. *Proceedings of the First Conference on the Mechanical Behavior of Salt*. Rockport, MA: Karl Distributors. 717-737.

Pudewills, A., Müller-Hoeppe, N. and Papp, R. (1995). Thermal and thermomechanical analyses for disposal in drifts of a repository in rock salt. *Nuclear Technology*. 112:79-88.

Pudewills, A. (1998). Thermomechanical analysis of the TSS experiment. In Franzen, T., Bergdahl, S.G. and Nordmark A., eds., *Proceedings of the International Conference on Underground Construction in Modern Infrastructure*. Stockholm, Sweden, Rotterdam: AA: Balkema. 317-323.

Senseny, P.E. (1986). Triaxial compression creep tests on salt from the Waste Isolation Pilot Plant. Sandia National Laboratories, SAND85-7261. Albuquerque, NM.

Van Sambeek, L.L., Fossum, A.F., Callahan, G. and Ratigan, J., (1993). Salt mechanics: empirical and theoretical development. In H. Kakihan, H.R. hardy, Jr., K. Koyodura and T. Hoshi, eds., *Proceedings of the Seventh International Symposium on Salt*. New York: Elsevier Science Publishers. 127-134.

Wawersik, W.R. and Hannum, D.W. (1979). Interim summary of Sandia creep experiments on rock salt from the WIPP study area, Southern New Mexico. Sandia National Laboratories, SAND79-0115. Albuquerque, NM.

ENVIRONMENTAL FLUID MECHANICS
ITS ROLE IN SOLVING PROBLEMS OF POLLUTION IN LAKES, RIVERS AND COASTAL WATERS

G.H. Jirka

University of Karlsruhe, Karlsruhe, Germany

Abstract. Environmental fluid mechanics has emerged as a strongly interdisciplinary research discipline over the last three decades. It is concerned with the understanding of the fluid motion and associated mass and heat transport processes that occur in the earth's hydrosphere and atmosphere on local and regional scales. In this article three examples drawn from the author's own research on environmental fluid mechanics are presented. The examples are: (i) gas transfer at the air-water interface of water bodies, (ii) turbulence structure and pollutant transport processes in shallow flows, and (iii) development of engineering expert systems for planning and prediction of pollutant releases into diverse water bodies.

1 Introduction

1.1 Definition and Raison d'Etre

Environmental fluid mechanics is concerned with the understanding of the fluid motions and associated mass and heat transport processes that occur in the earth's hydrosphere and atmosphere on local or regional scales. A particular emphasis within these scales – and in contrast to the yet larger domain of "geophysical fluid dynamics" – is the influence of these flows on and their interaction with man-made facilities and structures and their response to anthropogenic releases of mass and heat.

Thus, environmental fluid mechanics has a significant engineering dimension that plays part in the management, control and remediation of pollutant releases from municipal, industrial or agricultural sources into the natural environment. Its main impetus arose about thirty years ago with the advent of the environmental movement, marked most prominently by the first Earth Day in 1970.

Spurred on by a growing awareness of a creeping environmental degradation, punctuated by incidences of full-fledged ecological disasters, the fledgling environmental movement of the nineteen-sixties finally came to define itself in the public's mind on the occasion of the first Earth Day. Inexorably since then, this has served to foster discussion and raise public consciousness of society's impact on the natural environment, which, in turn, led to political initiatives and legislative actions to control that impact. As a consequence, in most industrial countries there have been substantial, sometimes spectacular, improvements in air, water, and land quality over this time period. Advances in waste control, abatement, disposal, and recycling technologies and practices have led to these improvements. As examples, rivers or lakes that before were considered unfishable and unswimmable have been restored to healthy habitats, and impaired air quality and acid rain events have been mitigated in many regions.

Nonetheless, as some of these problems have been solved. Others, hereforeto unrecognized or considered of lesser concern, keep coming to the forefront. Clearly, the maintenance of a sound environment quality in balance with limited natural resources will remain a continuing political and technical challenge for the future of modern societies. This challenge will be particularly acute for most of the developing world where the struggle against poverty and for economic survival often forces the complete abandonment of environmental quality concerns.

Major developments in environmental fluid mechanics have taken place during this period. These here led to methods and techniques that give scientists and engineers a solid causal linkage between the type and magnitude of a pollutant release – perhaps accidental or routine – and the resulting distribution in the air or water environment. This linkage, often in form of "predictive models" allows engineers or planners to devise plans for pollution prevention, for the clean-up or remediation of past pollution episodes and for granting permits for discharge releases at levels that do not harm the environment. Practically all industrial countries have enacted legislation for "environmental impact assessments" (EIS) for any major new public facility or industrial installation; predictive models that embody major elements of environmental fluid mechanics are practically always part of such EIS activities.

Environmental fluid mechanics is a strongly interdisciplinary subject. While its main element derives from fluid mechanics with the underlying transport and diffusion paradigms, it also has an intimate connection to chemical, biological and ecological aspects. The solution of environmental problems indeed requires, more often than not, the successful cooperation across these disciplines. This also represents a major challenge in the educational realm. While research in environmental fluid mechanics has been prospering over the last three decades, it has taken considerably longer for it to enter the classroom setting. Gradually, university curricula are requiring coursework in this discipline and some textbooks are being developed.

In this article three examples drawn from the author's own research on environmental fluid mechanics are presented. These are: (i) gas transfer at the air-water interface of water bodies, (ii) turbulence structure and pollutant transport processes in shallow flows, and (iii) development of expert systems for planning and prediction of pollutant releases into diverse water bodies.

The three examples are intended to illustrate problem types, research methodologies and transfer to practical engineering use. Thus, they demonstrate some of the substantial progress and state-of-the-art that has been achieved in recent years. Despite this progress major questions remain as a challenge. These research needs are reviewed in the following.

1.2 Outstanding Research Needs

The outstanding deficiency in environmental fluid mechanics has been, and continues to be, the transition from the microscopic continuum description of fluid flow – e.g., at the level of the Navier-Stokes equations – to the macroscopic continuum description, as dictated by the large scale dimensions of environmental applications. Because of this missing link that is common to other areas of engineering fluid mechanics, environmental fluid mechanics is replete with unsolved problems. Some of the important problems are:

Single-Phase Turbulent Flows: The problem of turbulent flows, i.e. linking the mean and fluctuating flow characteristics to the flow forcing functions and boundary conditions, is unsolved. All current treatments are empirical and statistical. Surprisingly, recent data suggest a substantial degree of orderly behavior ("coherent structures") superimposed on "random" turbulent flow. This finding emphasizes the need for improved flow stability analysis. Stability analysis for low Reynolds number non-turbulent flows need to be conducted to learn more about flow breakdown and turbulence generation. Similar stability techniques ought to be applied for high Reynolds number flows, using eddy viscosity analogies (despite their limitations) to learn more about the behavior of coherent structures and secondary flows that significantly affect engineering systems. Turbulence onset, generation and maintenance in stratified or rotating flow systems is another critical, but little understood, problem area.

Two-Phase Laminar or Turbulent Flows: The formulation of the dynamic equations for practically all two-phase disperse systems is empirical. The true dynamics remain a mystery. A key problem in hydraulics is solid particle transport. The elements of suspended load and bed load transport in turbulent flow are essentially unsolved. Present hydraulic "theories" are at great odds with available data. But even seemingly well behaved laminar flow systems, e.g. as suspension in a settling tank, show surprising and difficult behavior, such as wave-like front formations. High concentration mixtures that represent non-Newtonian fluid behavior, e.g. slurries, debris flow, mud slides, represent another difficulty. Gas-water mixture flows, as caused by the air entrainment at high-velocity hydraulic structures or by cavitating flows, pose yet further problems due to their thermodynamic complexities. Finally, low Reynolds number porous media flow (Darcy flow) is still waiting for satisfactory predictive explanation of its macroscopic properties, such as hydraulic conductivity.

Transport Phenomena: the present level of understanding is restricted in the area of transport processes for any materials contained in the water flow. Concepts, such as large scale eddy diffusivity, or hydrodynamic dispersion, still cannot be rigorously related to the actual flow field and solid matrix properties. Empiricism prevails, and further advances are needed as increasing amounts of work in hydraulic engineering practice are, in fact, devoted to the prediction of the fate and transport of materials in the environment or in engineered structure.

Interface Problems: The microscopic/macroscopic transition problem becomes especially severe and intractable at system boundaries, so-called interfaces. The air-water interface at the surface of a water body remains an enigma as far as the generation of wind waves, and their ultimate growth and instabilities, is concerned. Similarly, the water-sediment interface at a stream bed separating turbulent water flow from granular media behavior has not been successfully described. Systematic approaches that reconcile the large-scale macroscopic techniques to describe processes away from the interface with the microscopic processes directly at the interface are needed.

2 Gas Transfer at the Air-Water Interface

The transfer of gases across the air-water interface of environmental water bodies is a critical element in the natural geochemical cycling of materials as well as for the transport and transformation of pollutants in the environment. Problem areas range from such traditional concerns as oxygen transfer (to compensate for biochemical oxygen consumption), to reactive gases such as ammonia or carbon dioxide (important for the "greenhouse" effect), to toxic chemicals. In the latter category, for example, about one third of the "priority pollutants" as classified by the U.S. Environmental Protection Agency are volatilizing compounds for which the transfer to the atmosphere represents the major mechanism controlling aqueous concentrations and long-term effects.

All kinds of water bodies are affected by gas transfer processes. Civil/environmental engineers have commonly been concerned with reaeration processes in streams and rivers and have attempted to link the transfer rate to various stream parameters. Oceanographers and limnologists, on the other hand, have usually described gas transfer in the ocean or large lakes by considering wind as the forcing function. The estuarine/coastal environment with its diverse mixture of physical processes has received relatively little attention in past research despite its role as a major recipient of anthropogenic pollutants, either through import of materials by river inflows carrying upland industrial discharges or agricultural run-off, or through local aquatic disposal (routine or accidental) from urban, industrial, or navigation sources.

The proceedings of three recent international symposia on "Gas Transfer at Water Surfaces" (Brutsaert and Jirka, 1984: Wilhelms and Gulliver, 1991; Jähne and Monahan, 1995) and the review by Jähne and Haußecker (1998) give a yet wider overview on research progress and remaining problems in this area.

2.1 Physico-Chemical Processes in Surface Gas Transfer

The flux J (mass/time, area) of a gas into or out of the water column is commonly expressed as

$$J = K_L \left(C_s - C \right) \tag{1}$$

where C_s = gas saturation concentration (mass/volume), C = bulk mean concentration in water column, and K_L = overall gas transfer coefficient (velocity). C_s is related by Henry's law to the partial pressure p of the gas in the atmosphere

$$p = H_c C_s \tag{2}$$

in which H_c is Henry's law constant. For exotic gases, normally not present in the atmosphere, $C_s = 0$.

The overall transfer velocity K_L depends on molecular and turbulent processes in the air and water boundary layer close to the interface itself. The existence of admixtures (surfactants, microlayers) in the water phase is especially important here. For a clean interface the total K_L is usually represented by individual transfer velocities in liquid, k_L, and in gas, k_g, respectively, in a resistance-in-series model (Lewis and Whitman, 1924; Liss and Slater, 1974)

$$\frac{1}{K_L} = \frac{1}{k_L} + \frac{1}{H_c k_g} \tag{3}$$

From Eq.3 it is readily shown that the rate controlling component is determined by the ratio $k_L/H_o k_g$ (Coantic, 1980) If the ratio is small, then the water side controls; if it is larger, then the air side controls; and if it is of order unity, processes in both air and water are important. Estimates for k_L and k_g are obtained from materials that are clearly controlled by one side only, such as gases of very low solubility (e.g., O_2) for k_L, and direct water evaporation into the atmosphere for k_g. If such estimates are used and compared to known Henry's constants for volatilizing organic compounds the following situation emerges: Most toxic compounds in environmental conditions are water side controlled due to their low solubility (high H_c). Thus, the hydrodynamic effects on the liquid side of the interface dominate gas transfer. However, some compounds (halogenated hydrocarbons, e.g., hydrofluoric acid, and PCB (Mackay and Yuen, 1983) may be in the intermediate range so that the aerodynamic conditions above the surface are equally important for these.

The specification of the transfer velocity K_L (equal to k_L for water-side control, as will be assumed in the following) is a key element in any transport or water quality response model. Numerous predictive formulations for K_L have been proposed over the last three decades. These formulations fall usually into two classes:

1. Conceptual models (see review by Theofanous, 1984) in which a schematic mechanism for the near-surface interaction of molecular diffusion and the turbulent flow near the surface is assumed (e.g., in form of a film renewal or of roll cells). Apart from their physical simplification the main difficulty in these models is how to relate the assumed turbulent flow variables near the surface to the overall flow parameters.

2. Empirical models that represent a best fit (usually through multiple regression) between observed K_L values and some global variable, such as wind speed or mean water velocity. In general, these models lack physical validity and pose problems when applied outside their own database.

No detailed review of predictive model formulations is attempted here. Recent work has ascertained the following physical and chemical processes as controlling the transfer rate.

Near-Surface Turbulence: Detailed examinations of near-surface turbulence and their interaction with molecular processes at the interface have been given by Hunt (1984) and by Brumley and Jirka (1988). The principal turbulence sources are (a) mechanical shear energy away from the surface that generates turbulence, which then diffuses to the surface (e.g., channel flow, rivers, stirred grid experiments), (b) mechanical shear and turbulence generation directly at the surface (e.g., wind shear), and (c) turbulent buoyant convection (e.g., evaporative or radiative cooling effects of the surface). There have been various attempts to define a few simple common turbulent length and velocity scales in order to correlate transfer velocities K_L. But there appear to be such pronounced structural differences among these flows to make such attempts futile at this time.

Stream Turbulence. The major turbulence source in streams or rivers is shear at the bottom. This turbulence production is in equilibrium with local dissipation near the bottom. At higher levels, including the near-surface region, upward diffusion of turbulent energy balances the dissipation. A system that has qualitative similarities, but much easier to measure in considerable detail, is a stirred-grid tank experiment. Detailed measurements of the near-surface turbulence structure in such systems have been performed recently (Brumley and Jirka, 1987). No comparably detailed data are yet available for channel flow conditions. A whole host of gas transfer models, mostly developed in the civil/environmental engineering discipline, exists for stream flow situations. Most of these have highly divergent behavior at high or low flow conditions. A model comparison by St. John et al (1984) shows the impact of this model discrepancy on water quality decision making.

Wind Turbulence. From the viewpoint of gas transfer, the major effect of a steady wind shear acting on the water surface is the generation of a turbulent drift velocity profile. Additional effects relate to the stability of the surface itself leading to wave generation (discussed below), and the generation of large-scale inertial currents if basin boundaries are present. Commencing with the early work of Liss and Slater (1974) considerable effort has been expended on the development of correlations between wind velocity (or shear velocity) and K_L using both field data (see Broecker and Peng, 1982) and laboratory investigations on wind tunnels of all types (e.g., Jahne et al, 1987; Merlivat and Memery, 1983). A wide discrepancy of data can be seen depending on data source and tunnel type. In recent years wind-tunnel transfer studies also have become an accepted standard for the rate determination of toxic volatilizing organics for subsequent risk analysis using the obtained rate coefficients. Several theoretical models toward predicting wind controlled gas transfer are based on a flat-wall analogy to the turbulent boundary layer flow (e.g., O'Connor, 1983; Kitaigorodskii, 1984) with potential modifications for a wavy surface. These models contain a smooth wall (viscous) behavior at low wind speeds and a rough wall behavior at high wind speeds. Uncertainties exist in the transitional range of average wind speeds, which is important for environmental analyses.

Convective Turbulence. Following earlier studies by Hoover and Berkshire (1965), the effect of evaporation and condensation in the air layer was investigated by Liss et al. (1981) as regards transfer of oxygen in fresh water. They found little effect (only a minor one for condensation), which might not be surprising as oxygen is water-side controlled and any effect would be indirect through the stability of the air mass and its effect on shear transfer (drag). It is conjectured here that such effects may be stronger and more direct (a) for air-side controlled gases and (b) for salt water, where strong evaporation causes salt accumulation convective overturn and turbulence in the water column, hence directly affecting water-side controlled gases.

Combined Wind-Stream Turbulence. Many environmental situations are, in fact, strongly influenced by both wind and stream turbulence. Certainly, estuaries and coastal regions are in that range as well as run-of-the-river reservoirs or low-slope streams. The traditional model paradigms - transfer models based on wind effects for the oceanographic/climatological community versus models based on stream parameters for the civil/environmental engineering community - have to be expanded. Examination of recent field data and transition rules based on a near-surface energy dissipation approach have led Jirka and Brutsaert (1984) to conclude that most larger, low-slope rivers are actually strongly influenced, if not dominated, by wind effects. Correlation of gas transfer with stream parameters would lead, and often has led, to decidedly wrong predictions. Similarly, Kemp and Boynton (1980) have found that bottom shear generation from tidal currents may be the controlling mechanism in Chesapeake Bay rather than wind shear, as would be commonly assumed.

Surface Instabilities (Bubbles and Spray): The drastic enhancement of surface gas transfer by this mechanism has been pointed out in oceanographic literature (e.g. Su et al., 1984). Breaking waves lead to both bubble formation in water and spray formation in air which cause considerable amplification in the interfacial surface area as well as an increase in the turbulence level. As a consequence, a breaking wave ("whitecap") appears to act like a "vent" or "chimney" on gas transfer with a corresponding increase in the overall averaged transfer rate (Broecker, 1982). Average enhancements by a factor of 3 to 5 compared to smooth surface values have been consistently measured in several different wind-tunnel studies (Jahne et al., 1982). It should be pointed out, however, that these enhancements occur at relatively high wind speeds (<5 m/s). In any case, the mechanism for wave breaking (due to wind or as a wave instability) and bubble formation are far from resolved (e.g., Melville et al., 1985), even though certain scaling relationships on bubble size and distribution are beginning to emerge (Wu, 1988).

Surface Contamination and Microlayers: The presence of chemical films ("monolayers") due to pollution, or of natural organic surface layers, is another factor that seems to have considerable influence on gas transfer. The structure of such organic microlayers has been summarized by Hardy (1982) and by Lion (1984). Data from wind tunnel experiments and carefully controlled stirred-grid tank experiments by Asher and Pankow (1986) have demonstrated that K_L is very sensitive to the presence of interfacial films. In fact, it is very difficult to obtain truly "clean" laboratory conditions. Even the use of tap water in an experiment seems to have a detectable influence on transfer. This makes questionable all clean-surface assumptions used in the formulation of conceptual renewal models as discussed by Brumley and Jirka (1988).

Chemical Reactivity: The transfer velocity K_L is further modified if the gas is reactive with respect to species dissolved in the water column. Carbon dioxide, hydrogen sulfide, sulfur dioxide, and ammonia are all examples of environmentally important reactive gases. Generally these reactions have kinetic time scales much shorter than the hydrodynamic transfer time scale (e.g. Quinn and Otto, 1971) leading to flux augmentation, vis. an increased transfer velocity.

Laboratory experiments form a crucial tool for studying the details of the physico-chemical process that governs gas transfer from the very small molecularly dominated scales to large scale turbulent phenomena. This approach must include the development of large-scale, controllable experimental equipment, innovative and accurate measurement and visualization techniques to elucidate near-interface conditions, and attention to the non-uniformity of the actual fluvial, lacustrine, or coastal environment in which many of the physical sub-processes combine in a strongly non-linear fashion. The long-range goal of such research is the development of robust and reliable predictive techniques for environmental engineering practice. No immediate fulfillment for that goal is around the corner; there can be only a steady progression toward that goal through an incremental understanding of the numerous sub-processes that govern gas transfer.

2.2 Stirred-Grid Experiments

Early work at Cornell University concentrated on the development of a specially controlled environment in the form of a grid-stirred tank apparatus (see Fig. 1). This is perhaps the ideal environment for studying turbulent conditions that resemble those of riverine turbulence (with bottom generation and diffusion toward the surface). A complete mapping of the turbulent near-surface velocity field using a split hot-film anemometer was conducted (Brumley and Jirka, 1987). Earlier work by Hopfinger and Toly (1976) had shown that a vertically-oscillating grid can be used to produce horizontally near-homogeneous turbulence, and the root-mean-square horizontal velocity u' in a grid-stirred tank can be related to easily measured parameters such as grid stroke S, oscillation frequency f, and mesh size M,

$$u' = 0.25f\,S^{1.5}\,M^{1.5}\,y^{-1} \qquad\qquad (4)$$

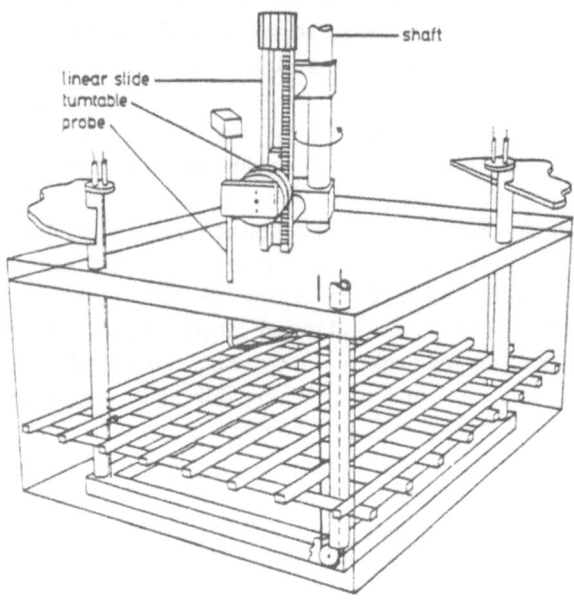

Fig. 1:   Schematic of grid-stirred tank used for near-surface turbulent velocity and gas concentration measurements at Cornell University. Tank dimensions are 50 cm square by 40 cm deep. The rotating probe assembly carries hot-film anemometry and a polarographic microprobe.

where y is the distance from a virtual origin approximately at the center of the grid stroke. In finite tank depth conditions with a free surface, Brumley and Jirka (1987) found the spatially averaged velocity fluctuations in the bulk region of a grid-stirred tank agreed well with Eq.4. But the turbulent structure is affected by the presence of the surface within a "surface-influenced layer", whose thickness is

$$L_\infty = 0.1 Z_s \qquad\qquad (5)$$

where L_∞ is the longitudinal integral length scale far from the surface, and Z_s is the distance from the surface to the virtual origin. This could be explained by the irrotational source theory of Hunt and Graham (1978). This theory assumes that the effect of a shear-free surface on a homogeneous turbulence field is the instantaneous superposition of an irrotational velocity field which cancels out the vertical velocity fluctuations there. Brumley and Jirka (1987) used the velocity at the surface $(y = Z_s)$ computed from Eq.4 as a velocity scale to characterize the different turbulence conditions, and called this velocity the Hopfinger-Toly velocity U_{HT}

$$U_{HT} = 0.25 f\, S^{1.5}\, M^{1.5}\, Z_s^{-1} \qquad\qquad (6)$$

The complete fluctuating turbulent velocity and length scale field above the stirred grid, and approaching the free surface is nicely described by these scales. As an example Fig.2 shows a the vertical and horizontal rms-velocity profiles, in comparison to the Eq.3 and to a modified equation accounting for the Hunt-Graham mechanism. Additional length scale and velocity spectral density

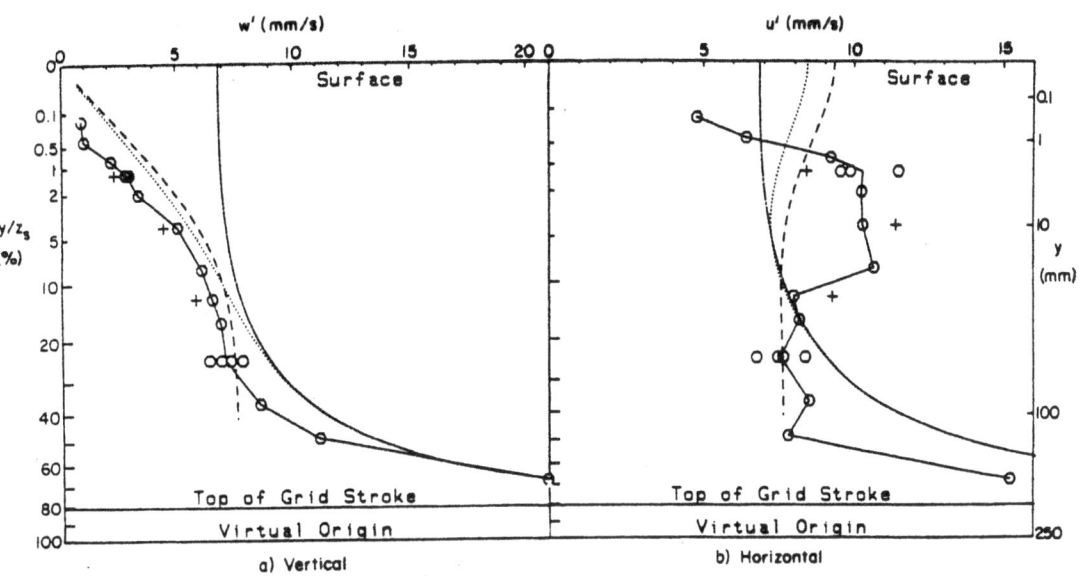

Fig. 2: Turbulent rms-velocity fluctuation profile in grid-stirred tank (from Brumley and Jirka, 1987). (a) Vertical, (b) horizontal velocity. --- Hopfinger-Toly relation (Eq. 3), --- Hunt-Graham profile, ... combined profile, o and + indicate experimental values. The vertical scale is distorted.

measurements, together with information from the gas concentration fluctuation measurements of Jirka and Ho (1990), led Brumley and Jirka (1989) to propose a classification of the near-surface

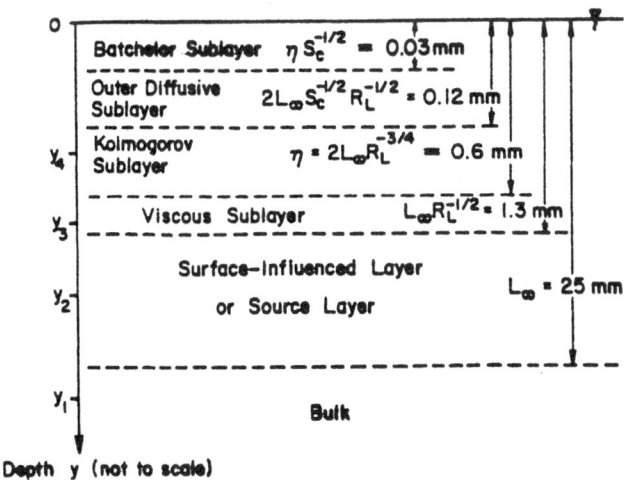

Fig. 3: Schematic sketch of the different hydrodynamic sublayers that influence mass transfer below the free surface. Dimensions are given for a typical stirred-grid experiment (from Brumley and Jirka, 1988).

region of a water body with turbulence diffusing from below into several distinct hydrodynamic layers (see Fig.3): a *surface-influenced layer* according to the Hunt-Graham mechanism, a *viscous sublayer* with viscous influence on the largest eddies of the turbulence spectrum, a *Kolmogorov sublayer* with viscosity influencing the smallest eddy scale, and two *diffusive sublayers* that measure the mass diffusive boundary layers within the turbulence structure and are smaller than the velocity boundary layers by the square-root of the Schmidt number $S_c = \nu/D$, in which ν is the fluid kinematic viscosity and D the molecular diffusivity of the gas. The dimensions given in Fig.3 are for typical conditions in these experiments and demonstrate vividly the small scales and multiplicity of scales inherent in these phenomena.

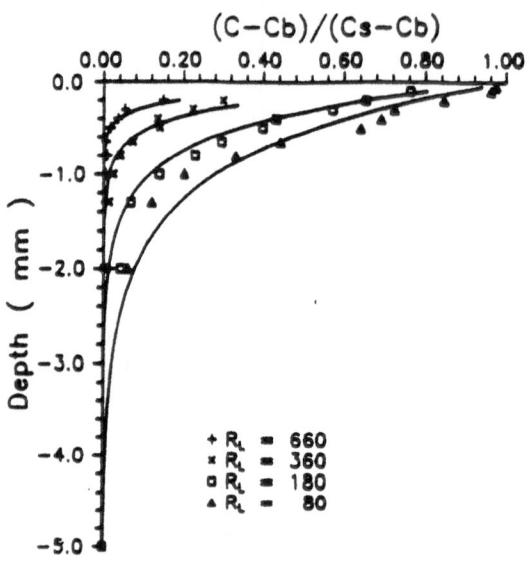

Fig. 4: Direct measurement of near-surface gas concentration (O_2) boundary layer with polarographic microprobe in grid-stirred tank (from Chu and Jirka, 1991).

Chu and Jirka (1991) used a polarographic oxygen microprobe (tip size 5 µm), in conjunction with the hot-film anemometer, to conduct more detailed concentration measurements and the first direct gas flux measurements (using the eddy correlation method) in such a turbulent system. Four transfer conditions, spanning a range of turbulent Reynolds number $R_L = 2U_{HT} L_\infty /\nu$ from 80 to 660. Multiple detailed measurements were performed at increasing distance below the surface. Fig.4 shows the concentration boundary layers measured for the four conditions. Fits of an exponential curve to these data (shown as solid lines in Fig.4) allow determination of the gas boundary layer thickness z_o (the 1/e depth). The profiles for high turbulence conditions have a steeper gradient than these for low turbulence conditions. This indicates that higher turbulence conditions clearly limit the thickness of the mean gas boundary layer. A comparison of the directly measured boundary layer thickness z_o, the computed Lewis-Whitman film thickness $\delta = D/K_L$ where K_L was obtained from separate measurements of the time change of the bulk concentration showed remarkable consistent behavior. Furthermore, they are also correlated to the diffusive sublayer thickness L_D defined by Brumley and Jirka (see Fig.3)

$$L_D = a L_\infty R_L^{-1/2} S_c^{-1/2} \tag{7}$$

in which a is a proportionality constant. Thus, z_0 (or δ) can be directly predicted from the knowledge of the near-surface turbulence properties, U_{HT} and L_∞. The diffusive sublayer thickness is consistent with the propositions of a large-eddy dominated transfer process (Fortescue and Pearson, 1967). These measurements constitute the first direct proof that support the conceptual framework of interfacial gas transfer that is controlled by molecular diffusion in a layer whose thickness is determined, however, by turbulent activity. They also support a large-eddy renewal process through actual observations in the transfer controlling regions. All earlier "proofs" had been indirect from overall mass balance (bulk changes). The direct data are also consistent with such indirect measurements. Fig. 5 shows the normalized transfer coefficient K_L/U_{HT} versus the turbulent Reynolds number R_L on a log-log scale. Furthermore, direct vertical flux measurements (correlations $w'c'$ in which w' = vertical velocity fluctuation and c' = concentration fluctuations) and measurements of the co-spectrum between w' and c' give additional support for the dominant transport role of the larger eddies, at least over the turbulent Reynolds number range of the experiment.

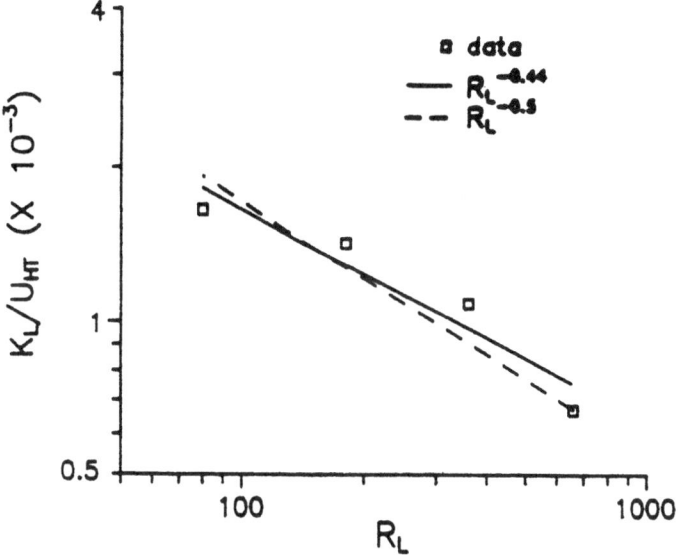

Fig. 5: Bulk gas transfer measurements in grid-stirred tank as a function of turbulent Reynolds number. Data support a large-eddy renewal model in this range (from Chu and Jirka, 1991).

As a next step in covering a wider class of hydrodynamic conditions, it seems obvious and imperative to attempt similar direct measurements at much higher Reynolds number conditions and in open channel flow. Clearly, such measurements will present substantial technical difficulties due to the yet smaller boundary layer thickness and the strong surface waviness in that range. But new measurement and laser visualization techniques (see Jähne and Haußecker, 1998) offer exiting prospects in this arena.

2.3 Experiments in a Tilting Wind-Water Tunnel

A Tilting Wind-Water Tunnel (TWWT) has been constructed in the DeFrees Hydraulics Laboratory at Cornell University as a facility dedicated primarily for gas transfer research, but also with some application to mixed layer research (lakes, reservoirs and ocean) and atmospheric boundary layer experiments. The TWWT (see Fig.6), which can be operated under water flow only, wind flow only, or combined, has a number of unique design features and sufficient size (20 m working length with a 1 m width and 0.8 m height cross-section) to make it ideally suited to a realistic simulation of riverine/estuarine/coastal gas transfer. The TWWT has sufficiently large flowrates (water flow 0.5 m^3/s, air flow 11.8 m^3/s) to provide turbulent Reynolds numbers $R_L = u_b^* h/v$, in which u_b^* = bottom shear velocity, h = water depth, and v = kinematic viscosity, up to 5,000, well above the transitional range of about 500 (Theofanous, 1984). It allows for sub- or supercritical water flow conditions with slope changes from -0.5% to +2.6%, and for reversible air flow. The entire system is corrosion-free (stainless steel or PVC in the return flow), an important feature for dissolved oxygen tests. Special turning vanes and vortex generators in the head-box provide for the generation of well-controlled fully turbulent inflow profiles, thus minimizing entrance effects that have plagued earlier facilities. Excellent operating experience has been acquired with perfect long-term stability, uniform turbulence conditions with minimal end-effects, and computer control of the entire operation.

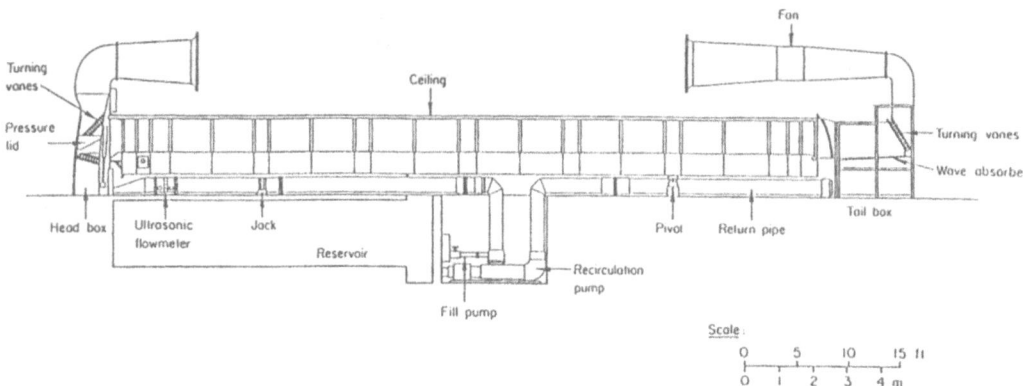

Fig. 6: General view of Tilting Wind-Water Tunnel (TWWT) at Cornell University.

Flow and Turbulence Structure: The detailed mapping of the mean flow properties and the turbulent structure in both water and air flow, with special emphasis on the conditions below and near the surface, was a major focus of the study. Two sets of separate benchmark experiments with water flow only and wind flow only, respectively, were followed by a series of combined wind/water flow experiments. A split-film hot-film anemometer was used to measure horizontal and vertical mean and fluctuating velocities. A capacitance wave probe recorded the associated surface wave fluctuations. More details can be found in Chu (1993).

The simple water and wind flow experiments exhibited well known logarithmic mean velocity profiles in accordance with earlier measurements. As for the turbulence properties in the water layer, Fig.7 shows the results for the (a) horizontal u' and (b) vertical velocity v' fluctuations, and (c) the energy dissipation rate ε, all as a function of elevation. Of particular interest, is the enhancement of the horizontal fluctuations and the damping of the vertical fluctuations near the water surface imposed by the kinematic constraint on the local eddies. This effect has been also observed in earlier studies with grid-generated turbulence (Brumley and Jirka, 1987). Different methods have been used to calculate the dissipation rate ε, giving consistent results. ε is a composite measure of local turbulent activity and a key parameter in some conceptual models for gas exchange.

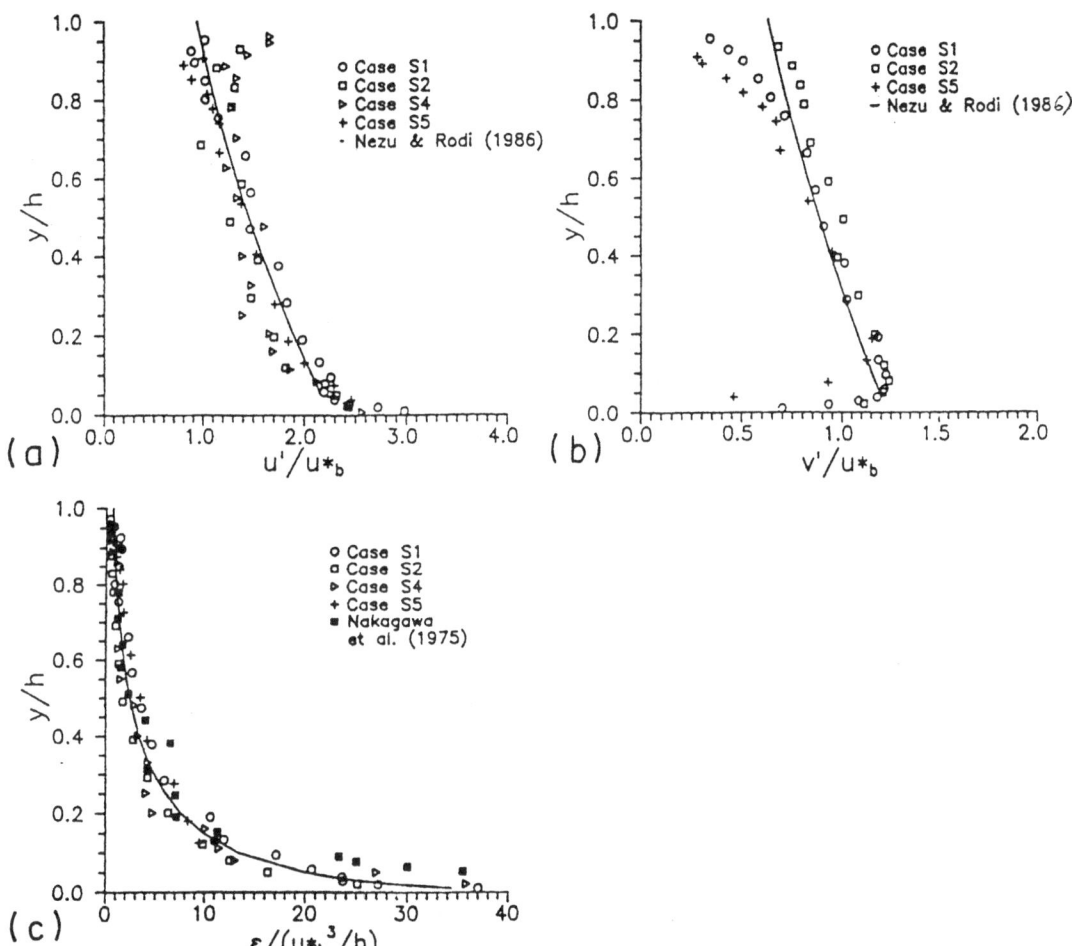

Fig. 7: Turbulent flow properties as a function of depth in pure open channel flow. a) Horizontal velocity fluctuations u′, b) vertical velocity fluctuations v′, and c) energy dissipation rate ε, all normalized by u⁎$_b$ and h.

In the combined wind-water mode the wind has little visible influence on the mean flow structure of the flowing water as shown in Fig.8a, but causes a significant enhancement of both horizontal (Fig.8b) and vertical (not shown here) velocity fluctuations. Great care must be taken to separate the fluctuating velocity fluctuations due to the wave induced (largely irrotational) velocity field from those due to the turbulent shear stress. The latter only will be responsible for turbulent gas transport. A spectral separation method proposed by Benilov et al. (1974) was used.

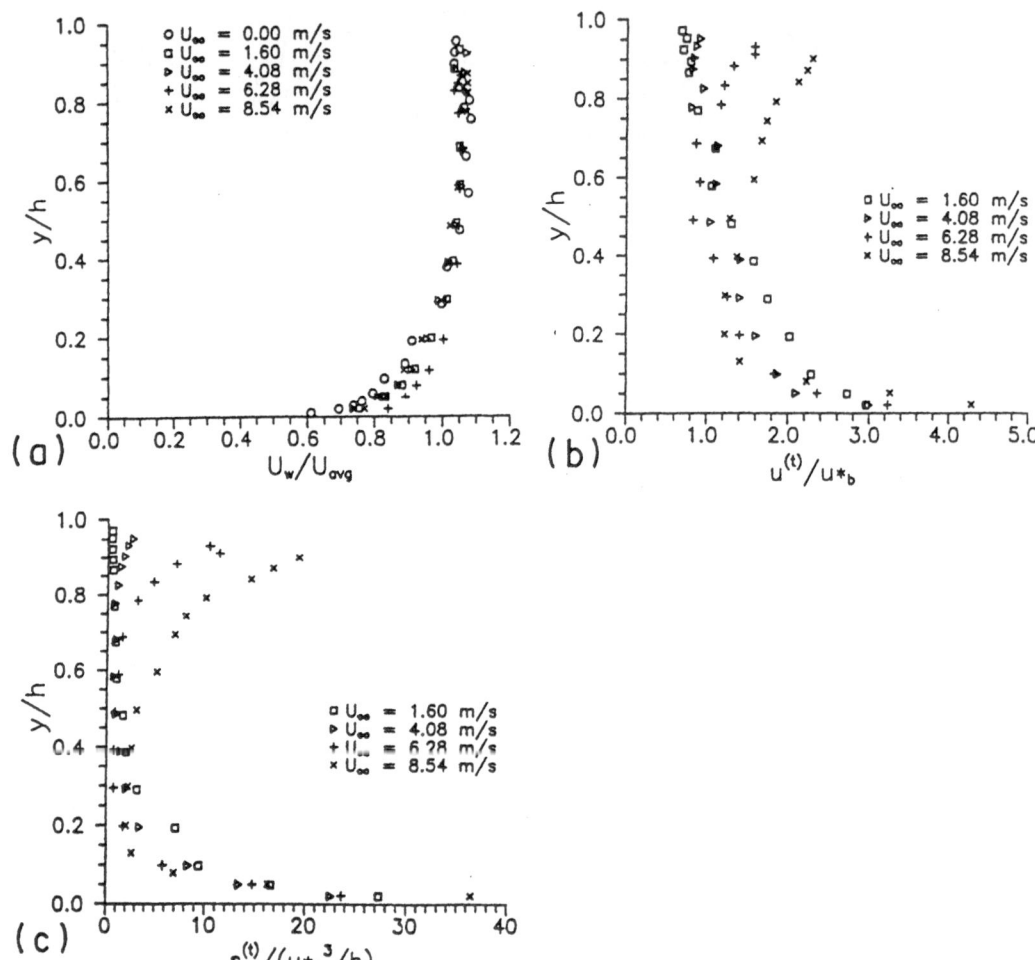

Fig. 8: Turbulent flow properties under combined wind/stream flow conditions. a) Normalized mean water velocity U_w, b) horizontal velocity fluctuations u′, and c) energy dissipation rate $\varepsilon^{(t)}$.

Gas Transfer: The gas transfer measurements were conducted with dissolved oxygen. The water was initially oxygen depleted by purging it with nitrogen in the underground reservoir.

Dissolved oxygen concentrations were measured by withdrawing a series of water samples at an upstream and downstream station in the TWWT, respectively, and subsequent Winkler titration analysis. The depth-averaged transfer coefficient K_L/h was then obtained from K_L/h = $(1/\Delta t) \ln(D_u/D_d)$ (Rathbun, 1988) in which Δt = travel time, and D_u and D_d are the oxygen deficits at the upstream and downstream station, respectively.

The measured transfer velocity K_L for pure open channel flow is shown in Fig.9 as a function of the bottom shear velocity u^*_b in comparison to other experiments. A reasonably large data scatter can be seen as is typical in such experiments, reflecting facility differences, additional parameters (e.g. surface impurities), and measurement inaccuracies. Also included in Fig.9 are the slopes indicated by the expectations of so-called large-eddy (1/2) and small-eddy (3/4) models of surface gas transfer. The experiments shown in the figure meet the condition, $R_L > 500$, so that the small-eddy model may be expected to be applicable (Theofanous, 1984). Indeed, this seems to be borne out by the data trend. Additional experiments (Moog and Jirka, 1999a; 1999b) give further support to the small-eddy model for large-scale laboratory - and presumably also field - conditions.

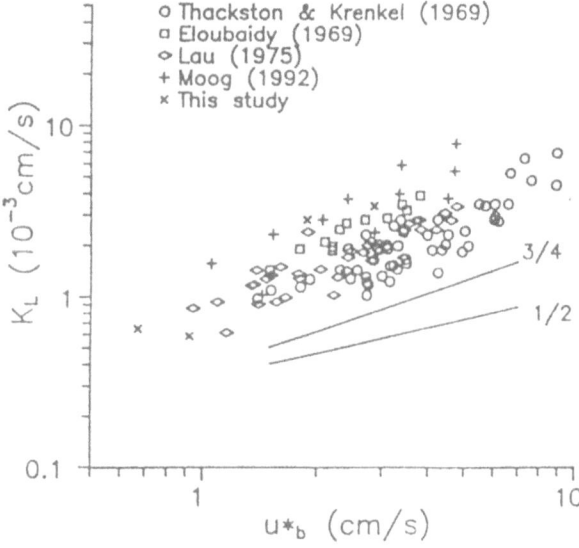

Fig. 9: Oxygen transfer velocity K_L for pure open channel flow as a function of the bottom shear velocity u^*_b.

In Fig.10, the experimental results for pure wind-controlled gas transfer are presented as a function of the wind-induced shear velocity u^*_a. A break in the data trend appears to take place at about $u^*_a = 15$ cm/s indicating a change in regime from an aerodynamically smooth to a rough water surface, respectively. For the low velocity regime, $K_L \sim u^*_a$, while for higher velocities, $K_L \sim u^{*2}_a$, in agreement with earlier studies (Deacon, 1977; Jahne et al., 1987).

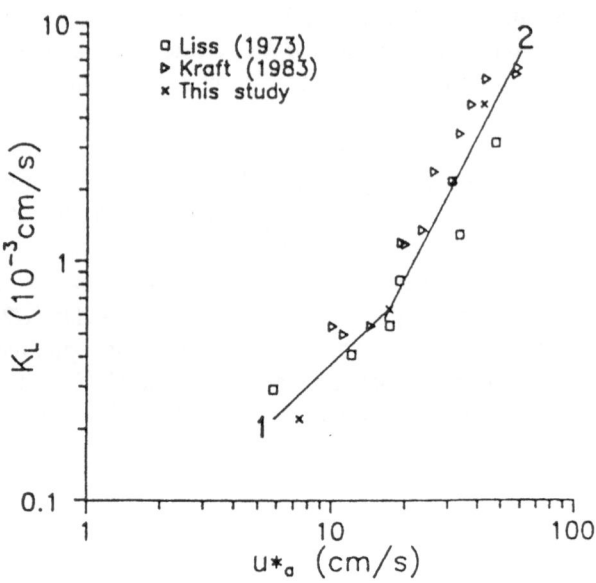

Fig. 10: Oxygen transfer velocity K_L or pure wind-driven flow as a function of the wind-induced-induced shear velocity u^*_a.

Finally, the measured transfer velocities for combined wind-stream turbulence cases are plotted in Fig.11 as a function of the wind shear velocity u^*_a. There are five experimental series, S1 to S5, with different stream flow conditions. The experimental series S0 refers to the base case of pure wind-driven conditions (see Fig.10). Fig.11 shows the strong enhancement of gas transfer with increasing wind speed, in particular for those cases that have weak stream-induced turbulence (such as S1, S3 and S5). Clearly, this is related to the increased turbulent near-surface activity that has been directly observed for those cases (see Fig.8 for case S1).

Predictive Model for Combined Wind/Stream Induced Gas Transfer: A mechanistic model for gas transfer is constructed by comparing the transfer rates caused by the bottom-shear and the wind-shear induced turbulence, respectively. For the bottom-shear induced transfer velocity K_{Lb}, an equation based on the small-eddy concept is compiled with recent highly accurate data from the TWWT experiments (see also Moog, 1995)

$$K_{Lb} = 7.38 \times 10^{-3} u^*_b R_L^{-1/4} \qquad (8)$$

where both K_{Lb} and u^*_b are in [cm/s]. The coefficient in the above equation is within $\pm 0.84 \times 10^{-3}$ (95% confidence interval). For the pure wind-induced turbulence condition, an approximation to the high-velocity, aerodynamically rough, regime is used

$$K_{Lw} = 2.09 \times 10^{-6} u^{*2}_a \qquad (9)$$

where both K_{Lw} and u^*_a are in [cm/s]. The coefficient is within $\pm 0.25 \times 10^{-6}$. A quadratic power dependence has been suggested in earlier approaches (e.g. Liss, 1973).

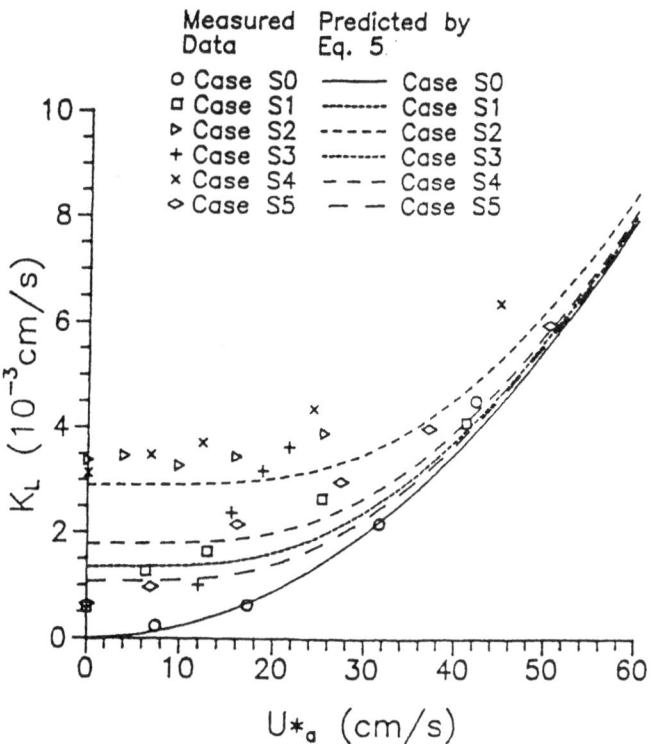

Fig. 11: Oxygen transfer velocity K_L as a function of the shear velocity u_b^* for different cases S1 to S5 of open channel flow. For reference, S0 is a pure wind flow case. Lines represent the predictions of Eq.11.

In combined wind/stream affected situations, the condition

$$K_{Lb} \approx K_{Lw} \tag{10}$$

presents a transition condition, while $K_b > K_{Lw}$ denotes bottom-shear control, and $K_{Lb} < K_{Lw}$ indicates wind-induced turbulence domination. A simple square-root matching, $K_L = (K^2_{Lb} + K^2_{Lw})^{1/2}$, between the two asymptotic regimes gives

$$K_L = 7,4 \times 10^{-3} \left(u_b^{*2} R_L^{-1/2} + 8,0 \times 10^{-8} u_a^{*4} \right)^{1/2} \tag{11}$$

where K_L, u_b^* and u_a^* are in [cm/s]. The curves given in Fig.11 represent, in fact, Eq.11 and show its ability to reproduce the observed data trends.

A comparison of Eq.11 to measured field conditions is provided in Fig.12 that includes two measurements from streams (Jirka and Brutsaert, 1984) and two from estuaries (Thames Survey Committee, 1964; Marino and Howarth, 1992). Despite substantial data scatter, Eq.11 gives acceptable predictions in both trend and magnitude of the transfer velocity.

It is interesting to use the transition condition, Eq.11, to demonstrate the relative impor-tance of <u>both</u> forcing functions in natural environments. Taking a wind velocity of 3 m/s as an average for many typical situations, yields an estimate for the wind-induced shear velocity in the water body, $u_a^* = 9,5$ cm/s, if typical wind stress coefficients are used (Chu, 1993).

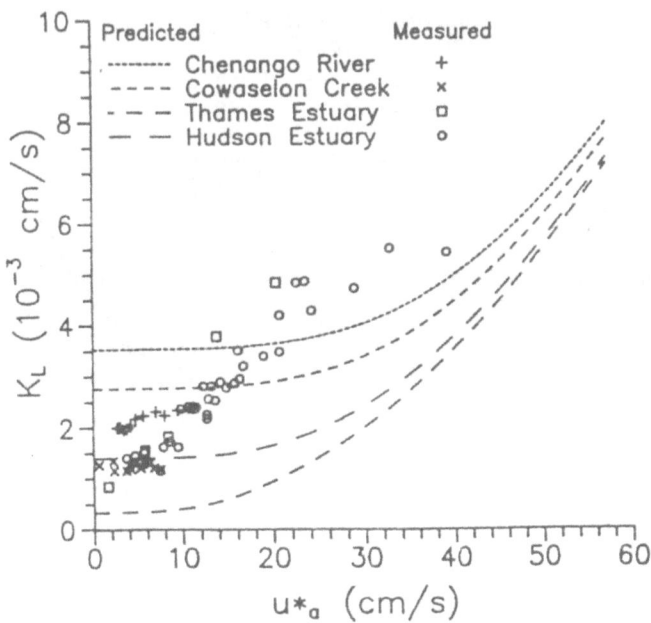

Fig. 12: Comparison of the predictive equation, Eq. 11 to field data for combined wind/stream flow conditions.

With a Darcy-Weisbach friction coefficient $f = 0.04$ typical for average water bodies, the transitional water velocity U_w corresponding to Eq.11 is given by $U_w\,(m/s) \approx 0.050\,h^{1/3}\,(m)$. For a <u>typical river ($h = 1$ m)</u>, the transitional $U_w = 0.07$ m/s. Hence, under average wind conditions if the water velocity exceeds 0.07 m/s the gas transfer is dominated by the stream conditions (bottom-induced turbulence), and if it is below 0.07 m/s it is dominated by wind conditions. Therefore, in shallow estuaries or coastal lagoons both wind- and bottom-induced turbulence (due to tidal currents) will dominate at different stages of the tidal cycle. Similarly, slow flowing streams (lake, reservoir) are more affected by wind conditions than by stream conditions. Assuming a <u>deep water body ($h = 10$ m)</u>, the transitional water velocity is somewhat larger, $U_w = 0.11$ m/s while actual flow velocities are typically lower. Thus, the average wind dominates over a wider range of water current velocities. This has implications on gas transfer in run-of-the-river impoundments in which the depth is increased while the mean velocity is decreased relative to the natural river conditions.

Clearly, these results show that the traditional model paradigms - transfer models based on wind effects used by the oceanographic/climatological community versus models based on stream parameters used by the civil/environmental engineering community - must be abandoned. Detailed attention should be paid to the actual time-dependent forcing functions, if accurate predictions of gas transfer in water quality models or coupled air/water transport models (as used in climate simulations) are desired.

3 Turbulence Structure and Transport Processes

Shallow flows are common in nature. We define shallow flows as predominantly horizontal flows in a fluid domain for which the two horizontal dimensions greatly exceed the vertical dimension. Flows in wide rivers, in estuaries and coastal waters, in shallow lakes or in the upper mixed layer of deep stratified lakes or reservoirs are important examples of hydraulic or environmental engineering concern. Yet larger scale geophysical flows in the shallow atmosphere surrounding the earth or in ocean basins - in both cases with or without the influence of stratification that further enhances layer formation - are also in that category.

Attention is limited herein to predominantly one-dimensional shallow flows (without significant curvature effects as in river bends) that are at scales below the Rossby radius (thus, free of Coriolis force influences) and that are fully turbulent. The turbulence condition is measured by a depth Reynolds number $Re_h = UH/v$, in which U is the characteristic velocity, H the flow depth and v the kinematic viscosity, that is sufficiently greater than 10^3. This is readily satisfied by the real-world flow, but may pose some difficulties in laboratory experimentation. The base flow is governed by wall turbulence, produced by the shear effect at the solid bottom. The structure of this turbulence is 3-D produced by ejection and sweep events near the solid boundary and characterized in the mean by a logarithmic-law velocity profile. Some types of coherent turbulent structures appear to be present in this flow (for a review see Nezu and Nakagawa, 1993) but the length scale of these vortical elements is of the order of or less than the water depth and their axes are aligned in the mean flow direction. Thus, these structures still represent 3-D turbulence quite distinct from the features discussed in the following.

Whenever such shallow flows - that may be uniform and wide in both horizontal directions - are subjected to localized or distributed disturbances they tend to undergo internal oscillations that grow into large-scale instabilities characterized by coherent structures. This is exemplified by Fig. 13 in the field or laboratory. Large-scale vortical structures are generated downstream of relatively small island obstacles. These structures are essentially two-dimensional, i.e. their horizontal extent ℓ is much larger than the depth H, $\ell/H >> 1$, they have vertically aligned vorticity vectors and the distribution of turbulent kinetic energy among smaller scales (but still larger than H) is governed by the laws of "2-D turbulence".

In the following, a classification is developed for the generation mechanisms that produce large-scale 2-D coherent structures in shallow flows. Furthermore, recent results for two types of transversely sheared flows, namely the shallow wake and the shallow jet, are presented elucidating the role of the coherent structures in affecting momentum and mass transfer (mixing).

3.1 Classification of 2-D Coherent Structures in Shallow Flows

2-D coherent structures (2DCS) are defined herein as connected, large-scale turbulent fluid masses that extend uniformly over the full water depth and contain a phase-correlated vorticity (with the exception of a thin near-bottom boundary layer). This definition is an adaptation of Hussain's (1983) definition for general (3-D) coherent structures.

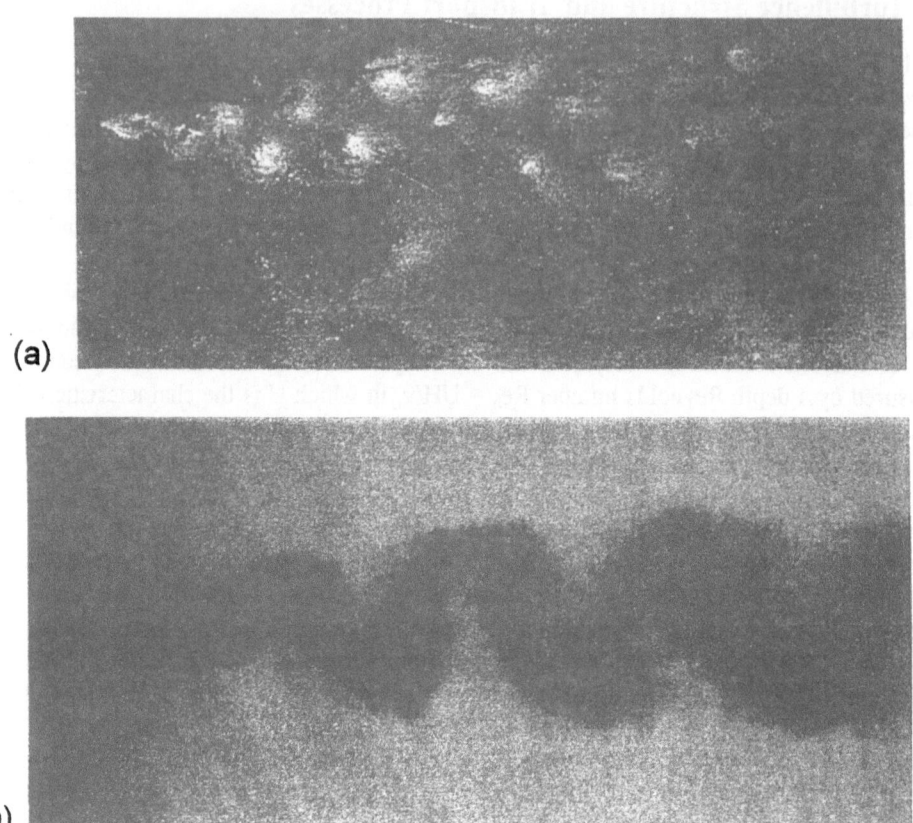

Fig. 13: Shallow flow containing large-scale coherent structures consisting of 2-D vortical elements. a) Flow downstream of small islands on Great Barrier Reef, water depth 10 to 20 m, horizontal extent several tens of km (photograph courtesy of E. Wolanski), b) laboratory simulation, water depth 3 cm, cylinder diameter 17 cm, horizontal extent 3.5 m (from Chen and Jirka, 1995). Flow is from left ot right.

The vorticity contained in 2DCS emanates from the initial transverse shear that has been imparted on these flows during their initial generation. Accordingly, Jirka (1998) has defined three types of **generation** mechanisms for 2DCS, listed in order of their strength:

Type A: Topographical forcing: This is the most severe generation mechanism in which topographic features (islands, headlands, jetties, groynes etc.) lead to local flow separation, formation of an intense transverse shear layer and return velocities in the lee of the feature. An example of this flow type is given in Fig. 14, in which the growth of 2DCS within the shear layer that starts at the separation point can be clearly seen.

Type B: Internal transverse shear instabilities: Here velocity variations in the transverse directions that exist in the shallow flow domain give rise to the gradual growth of 2DCS. Such lateral velocity variations can be caused by a number of causes: due to source flows representing fluxes of momentum excess or deficit (shallow jets, shallow mixing layers, shallow wakes) or due to gradual topography changes or roughness distributions (e.g. flow in compound channels). Fig. 15 shows a shallow jet entering a laboratory basin in which the flow takes on a meandering character.

Fig. 14: Tidal inflow into shallow lagoon shown by thermal imagery. Cooler (dark) ocean water enters under bridge on top left, warm (light) water is blocked by causeway on bottom left. A series of large 2-D coherent structures develops in shear layer.

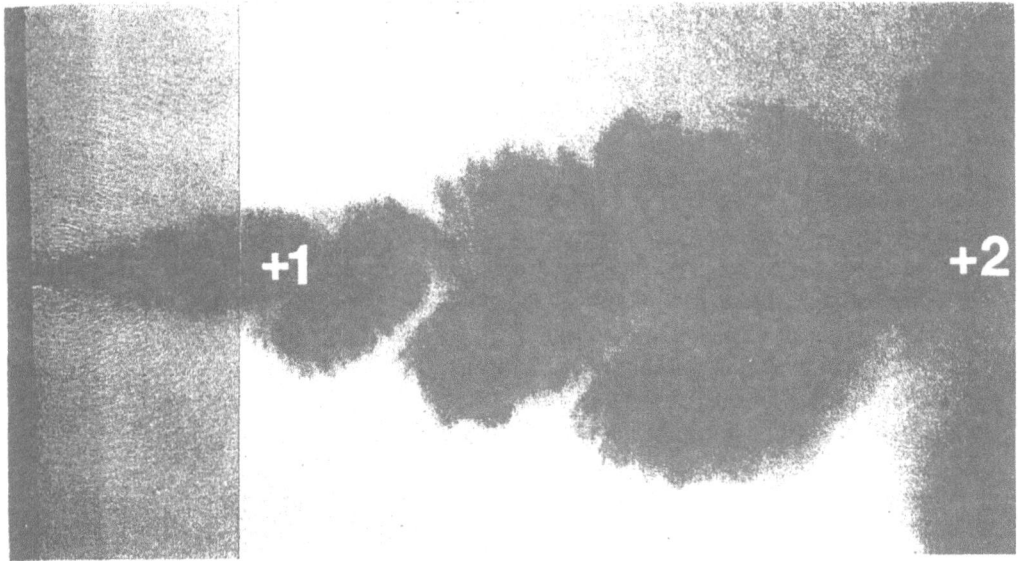

Fig. 15: Shallow jet (exit width B = 1 cm, H = 2.5 cm, Re = $U_oB/\nu \approx 10^4$) entering large flow basin (width of view about 3.5 m) (after Chen and Jirka, 1998).

Type C: Secondary instabilities of base flow: This is the weakest type of generating mechanism and experimental evidence is still limited. As remarked earlier, the nominal base flow is a uniform wide channel flow that is vertically sheared and contains a 3-D turbulence structure controlled by the bottom boundary layer. Slight imbalances in this flow process can lead to a wholesale redistribution of the momentum exchange processes at the bottom boundary, including as an extreme case separation of the bottom boundary layer. The distortion of the vortex lines caused by these flow imbalances lead ultimately to 2DCS. Contributing factors may be localized roughness zones or geometrical elements (underwater obstacles). The studies of Lloyd and Stansby (1997b) on submerged islands point in that direction (though there is also some connection to Type A mechanisms in their case). Gradual decelerations in the base flow (spatial or temporal, vis. tidal oscillations) can also lead to a breakdown of the base flow into 2DCS. The simulations by van Hijst et al. (1996) on cell formation in a shallow water tank seem to be examples of that. In either instance, the transverse momentum exchange induced by these flow patterns may explain the perplexingly high friction factors (Darcy-Weisbach coefficients) that have been found necessary when hindcasting numerical model results for flows in very wide open channels.

Whenever dealing with these generating mechanisms (especially Types A and B) it must be recognized that the generation of 2DCS always necessitated some travel time or convective distance from the origin of generation. A detailed analysis by Dracos et al. (1992) for the shallow jet and by Uijttewal and Tukker (1997) for the mixing layer actually shows three regions of development. In the "near-field" ($x/H \leq 1$) the transverse shear flow is 2-D in the mean, but contains highly 3-D small scale turbulence. The "middle-field" has a significant interaction of the mean and turbulent flow with bottom and surface, producing a strongly 3-D mean flow and 3-D turbulence. In the "far-field" ($x/H \geq 10$) the turbulence structures have grown to sizes greater than the water depth and for purely kinematic reasons these large eddies have vertical axes and hence 2-D character. Also, the mean flow, whose transverse scale is of the same order as the large turbulence eddies, becomes 2-D.

Following their generation the **growth** of 2DCS is governed by various processes. As they constitute turbulent vortical elements they grow by entrainment or engulfment of outside non- or less-turbulent fluid. Also pairing of separate structures leads to larger structures as observed in shallow jets (Dracos et al., 1992). In this fashion the 2DCS grow larger over time. The distribution of turbulent kinetic energy over different scales of the 2-D eddies is governed by the tenets of "2-D turbulence" theory (e.g. Kraichnan, 1967). Following the concept of the "enstrophy" cascade the flux of rotational momentum is constant among the different eddy scales leading to a k^{-3} distribution of turbulent kinetic energy in which k is the wavenumber. In 2-D turbulence there also exists the possibility of an inverse energy transfer in which energy is transferred from smaller to large scales. The vortex stretching mechanism that is the key attribute to usual 3-D turbulence by which energy is transferred to smaller and smaller scales until it is dissipated by viscosity does not exist in 2-D turbulence conditions.

The major mechanism that leads to the final **decay** of 2 DCS in shallow flows is the bottom friction at the base of the vortical elements. This friction is described by a shear stress τ_b = $\rho c_f U^2/2$ in which ρ is the fluid density and c_f a quadratic law friction coefficient. Typically, $c_f \approx 0.005$ (field) to 0.01 (laboratory). One can readily show that for an eddy size $\ell_{max} \approx 2H/c_f$ the eddy looses its rotational energy during one turnover. Thus, ℓ_{max} can be considered as the maximum eddy size that 2DCS can obtain in shallow flows. With the above values of c_f, this corresponds to a relative size $\ell_{max}/H \approx 0(100)$.

3.2 Methods of Investigation

While the occurrence and the gross features of shallow flow instabilities are well established in hydraulic and environmental flows - mostly from aerial photographs - it is difficult to observe them in the field in a systematic fashion in order to ascertain their dynamic properties. This is best accomplished by laboratory experiments supported by analytical methods and numerical simulations.

Experiments on shallow flows require wide laboratory basins or "water tables" of sufficient size and with good flow control. A number of laboratories world-wide have used such installations to successfully simulate different flow features. This includes ETH Zurich (Giger et al., 1991), Cornell University (Chen and Jirka, 1995), University of Manchester (Lloyd and Stansby, 1997a,b) and Delft Technological University (Uijttewal and Tukker, 1997). Another shallow water basin has recently been installed at the University of Karlsruhe. Ideally, the basin dimensions should be at least 4 m wide and 10 m long in order to be free of boundary effects on the 2DCS. Also, Reynolds numbers Re_h exceeding 2000 can be readily maintained in such installations providing a fully turbulent base flow. Experiments with smaller size facilities (e.g. Chu and Babarutsi, 1988, on the mixing layer) may provide some qualitative insight into these processes. Their results, however, do deviate in quantitative terms from the large size installations and, supposedly, from the field behavior. Modern observational techniques are dye concentration measurements by means of planar laser-induced fluorescence (LIF) and velocity measurements by LDA or by PIV-field methods.

Stability analyses of the shallow water flows provide useful insight into the mechanisms for onset and growth of the 2DCS. For that purpose the depth-integrated momentum equations are linearized, leading to a form of Orr-Sommerfeld equations containing bottom shear stress terms. This approach was first suggested by the work of Chu and co-workers (e.g. Chu et al., 1983) and later extended by Chen and Jirka (1997, 1998) for wakes and jets, respectively. Despite the limitations of this approach, such as linearization, assumption of purely parallel (rather than expanding) flow and simple eddy viscosity formulation for the lateral shear stresses, it yields important diagnostic information on the conditions for growth and/or suppression of the instabilities that lead to 2DCS. As an example, Chen and Jirka (1997) have shown that the two forms of growth mechanisms that occur in a shallow wake, namely absolute and convective instabilities, correspond to different wake structures (Type A and Type B, as discussed in the following section) that can be observed experimentally. It appears that many additional questions, such on the effects of lateral boundaries or the role of slight transverse shear, can be usefully explored with this technique.

Numerical simulation models are, of course, another promising method for the understanding of these flows. Since the Reynolds number domain is outside the realm of current direct numerical simulation (DNS) techniques, some form of turbulence closure method needs to be employed. Furthermore, though the flows are 2-D in their gross feature, in their detail they are, of course, 3-D. While a depth-integrated model offers much computational simplicity, it may suppress important details on the vortex generation (i.e. the aforementioned progression from the "near-" to the "middle-" to the "far-field" of the initial shear layer).

These questions need to be explored in the development of appropriate models. Ultimately, the large-eddy-simulation (LES) technique appears most obviously suited for a realistic simulation of these flows, but has not yet been implemented. Strongly forced (Type A) flows have been modeled with a k-ε technique by Lloyd and Stansby (1997a) showing reasonably good, although not fully satisfactory, agreement with observations on vortex-street wake flows. Whether a k-ε model would be successful for predicting more moderate Type B flows is quite doubtful. Only LES models appear appropriate for that.

3.3 Shallow Wakes

A few salient results from recent work are presented here to show the different types of 2DCS that occur in the shallow wake and their role in the mass exchange of advected material.

The behavior of shallow wake flow in the lee of an obstacle of diameter or width D is controlled by the wake parameter

$$S = c_f \frac{D}{H} \tag{12}$$

(Ingram and Chu, 1987; Chen and Jirka, 1995). As shown in Fig. 16, Chen and Jirka have classified three forms of wake patterns: i) the vortex street wake with unsteady separation at the cylinder, $S < 0.2$, ii) the unsteady bubble wake with a recirculating bubble attached to the cylinder that becomes unstable with sinuous oscillations further downstream, $0.2 \leq S \geq 0.5$, and iii) the steady bubble wake that shows no oscillations, $S \geq 0.5$. In essence, the vortex street pattern represents a Type A generation of 2DCS due to the flow separation at the cylinder periphery, while the unsteady bubble wake is a Type B generation due to the transverse shear imposed by the wake momentum deficit.

The wake behavior has been found to be independent of both depth Reynolds number Re_h - provided that $Re_h \geq 1500$ - as well as cylinder Reynolds number $Re_d = UD/\nu$ that are noted in the figure caption. The behavior seems Reynolds-invariant. The vortex street pattern (see also Fig. 13) looks remarkably similar to that for a von Karman vortex street in the laminar/turbulent transition of unbounded cylinder wakes, in the range $50 < Re_d < 300$. Secondary, spanwise instabilities rapidly destroy that vortex street pattern for higher Re_d in the unbounded case. In the shallow, bounded case, however, these primary instabilities remain the dominant ones because of the kinematic constraint and the flow becomes governed by these.

Whenever the shallow wake parameter S becomes large, bottom friction suppresses the vortex shedding process and any subsequent transverse instabilities leading to full stabilization.

Qualitatively similar observations can be made for wakes produced by solid plate-like obstacles. With the purpose of eliminating the initial flow separation (and thus, Type A generation) Chen and Jirka also employed a porous plate device that allows for some through-flow. In that case the vortex street pattern never occurred, but as shown in Fig. 17 only the unsteady bubble wake (Type B generation) or the steady wake are possible. The limiting value between two patterns was about $S \approx 0.19$ for the particular porosity (about 50%) of the porous plate. These results support the selectivity of the flow patterns as a consequence of type of generation as well as the shallow wake parameter S.

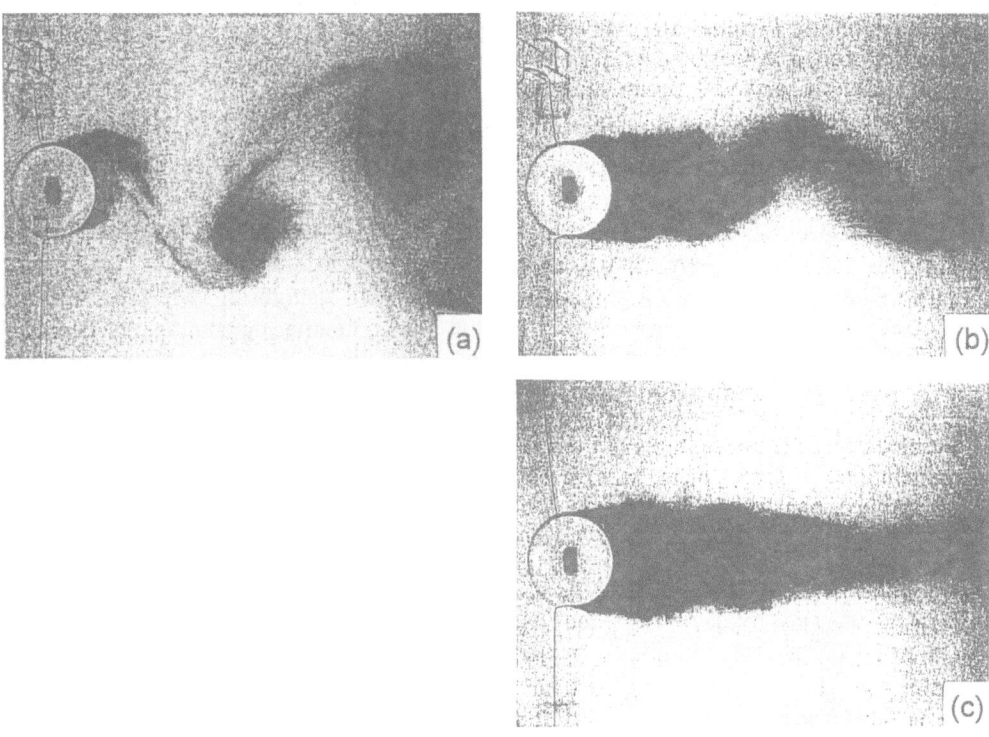

Fig. 16: Flow patterns of shallow wakes produced by a cylinder obstacle. a) Vortex street patterns (S = 0.19, Re_h = 5900, Re_d = 183000), b) unsteady bubble wake (S = 0.34, Re_h = 2600, Re_d = 115000), and c) steady bubble wake (S = 0.53, Re_h = 1800, Re_d = 112000) (after Chen and Jirka, 1995).

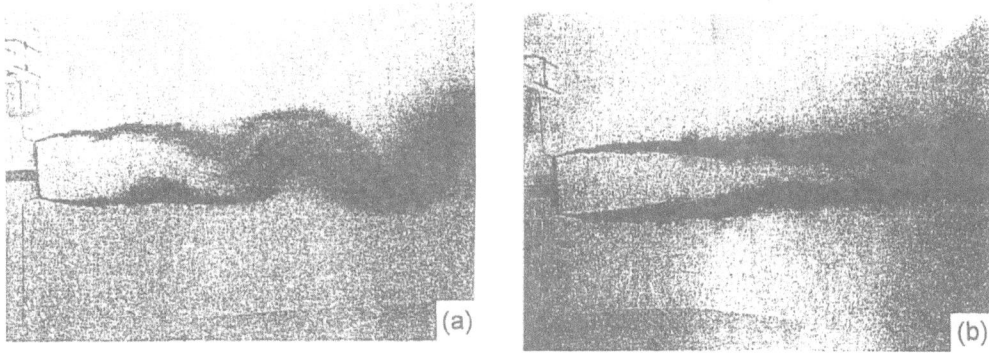

Fig. 17: Shallow wake patterns produced by porous plate device that allows for some ambient flow through obstacle. a) Unsteady bubble wake (S = 0.16, Re_h = 4850, Re_d = 121000), and b) steady bubble wake (S = 0.26, Re_h = 1780, Re_d = 55000) (after Chen and Jirka, 1995).

The different types of turbulent shallow wakes obviously have greatly differing mixing characteristics for any material, such as instantaneous or continuous discharge of pollutants that enter the wake region. As an example Fig. 18 shows the concentration pattern obtained by a planar LIF measurement in a cylinder wake that is near the transition between the vortex street and unsteady bubble flow pattern ($S \approx 0.2$). The dye that has been continuously injected at the two cylinder shoulders has been concentrated into vortical blobs while the concentration in-between is much lower. In fact, systematic measurements (Chen and Jirka, 1991) have shown that these instantaneous concentration peaks are about 4 to 6 times larger than the time-averaged maxima that occur at a particular downstream distance. This behavior has substantial bearing on the interpretation of pollutant data monitoring as may be employed by environmental enforcement agencies. Signals at a fixed monitoring point exhibit considerable intermittency. On the other hand, there may be very high exposure levels for organisms that are advected along in these concentrated blobs and that may have a sustained lifetime until they disintegrate.

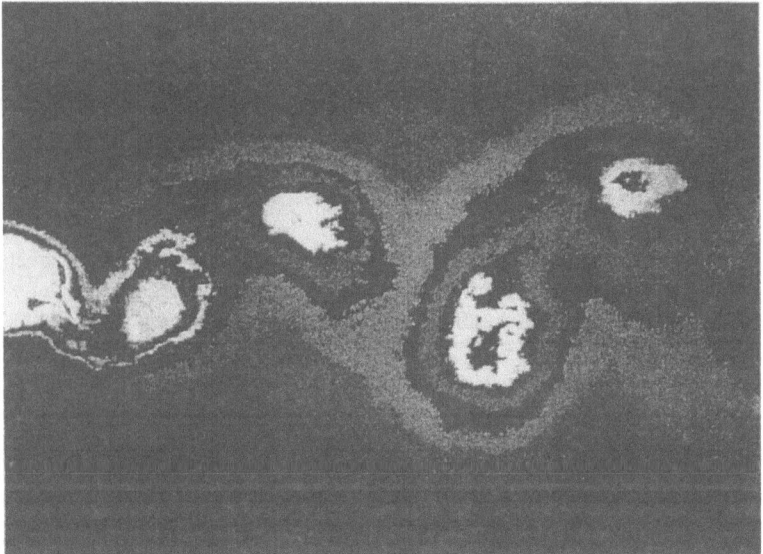

Fig. 18: Instantaneous view of concentrated dye pattern in a shallow cylinder wake, $S \approx 0.2$, as measured by planar LIF techniques. Cylinder is located at left edge of picture.

The existence of attached bubble flows for wakes with high S values has further influences on pollutant trapping in the lee of islands. The flushing time for such trapped pollutants (stemming for example from a treatment plant located on the island) can be a factor of 10 larger than for shallow wake flows with the more active vortex street pattern (Chen and Jirka, 1991; Lloyd and Stansby, 1997a).

Considerably weaker types of instabilities have been found for headland wakes in which because of the symmetry imposed by the shoreline the more vigorous sinuous oscillations are suppressed and only varicose instabilities are possible (MacDonald and Jirka, 1997).

3.4 Shallow Jets

The shallow jet (see Fig. 15) has been investigated through detailed velocity measurements (Dracos et al., 1992) and LIF concentration measurements (Chen and Jirka, 1998). These results have been summarized by Jirka (1994).

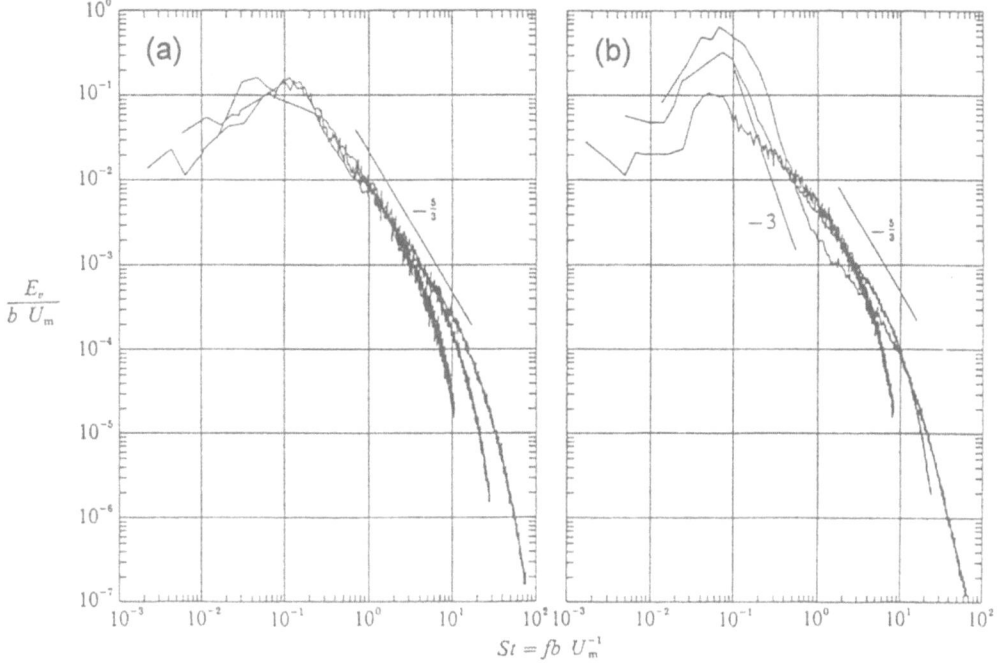

Fig. 19: One-dimensional normalized spectra of turbulent fluctuations of the transverse velocity. a) Spectra for three locations in the near/middle field, and b) spectra for three locations in the far field with increasing distance x/H.

As mentioned above a shallow jet enters its "far-field" at a distance, $x/H \approx 10$, beyond which it takes on its distinct meandering character with alternating vortical elements. Fig. 17 shows examples for one-dimensional energy spectra of the transverse velocity component v' measured on the jet axis. The spectra are scaled with the local centerline velocity U_m and half-width b; the non-dimensional frequency f abscissa can therefore be seen as a Strouhal number defined as $St = fb/U_m$. Fig. 19a shows near/middle field spectra. Their overlapping is indicative of similarity. There is a distinct energy peak around $St = 0.10$. This value agrees with some observations of weakly energetic large-scale structures in these types of flows (e.g.

Thomas and Goldschmidt, 1986). These three-dimensional structures, however, break down through transverse instabilities and undergo a vortex stretching mechanism thereby, transferring their energy to smaller scales. A universal equilibrium subrange with a -5/3 wavenumber dependence, typical for three-dimensional cascading turbulent flow, is therefore observed at higher wavenumbers.

Power spectra in the far field (Fig. 19b) behave quite differently. Here dynamic self-similarity no longer exists. Rather the maxima in the non-dimensional energy density become more pronounced with increasing jet distance. At the same time, a range develops in which the energy transfer follows a -3 wavenumber dependence. Such dependence is consistent with a quasi-two-dimensional turbulence characterized by an enstrophy cascade (Batchelor, 1969). The increase of energy at the low wavenumbers is associated with a depletion of the energy content at higher wavenumbers. At even larger wavenumbers the energy transfer gradually relaxes back to that for three-dimensional turbulence. The shapes of the spectra suggest that some energy is extracted from the inertial subrange of the spectrum at the location where the enstrophy cascade begins and is transferred in an inverse cascade back towards the peak which increases in magnitude.

The Strouhal number of the peak, St = 0.08, as obtained from the spectra, can also be observed by a variety of other means, such as autocorrelation functions and counting of the passage of the visible vortical structures, all of which agree closely (Dracos et al.). As the dimensional frequency is decreasing with increasing distance there must be a loss in the number of vortical elements. Indeed, this occurs through a pairing mechanism of similarly rotating elements on a given jet side.

The power density spectral distributions for the concentration fluctuations are shown in Fig. 20 for the two jet locations, labeled "1" and "2" in Fig 15. Again a transition from a -5/3 frequency dependence to a -3 dependence at large distances in the shallow jet is apparent. Thus, there is a consistent indication that the meandering shallow jet with its 2DCS is governed by the energy transfer characteristics of 2-D turbulence.

This also has repercussions on the mean and rms concentration fields as shown in Fig. 21. The lateral distance y is normalized as y/x which not only shows the self-similarity of the mean profiles (Fig. 21a), but also indicates a linear concentration half-width, $b_c/x = 0.17$, that is considerably larger than the velocity half-width spreading coefficient, $b/x = 0.10$ (Giger et al., 1991). In fact, the dispersion ratio $\lambda = b_c/b = 1.7$ greatly exceeds that for the usual unbounded plane jet, $\lambda \cong 1.35$. This aspect seems to be an indication of the occasional passage of larger vortical elements that contain high concentrations at or beyond the mean jet periphery. The rms-profile (Fig. 21b) shows maximum activity at a location equal to the concentration width b_c. Again compared to the unbounded case (Davies et al., 1975) there seems to be an increasing fluctuation activity caused by these intermittent 2DCS with increasing distance along the shallow jet.

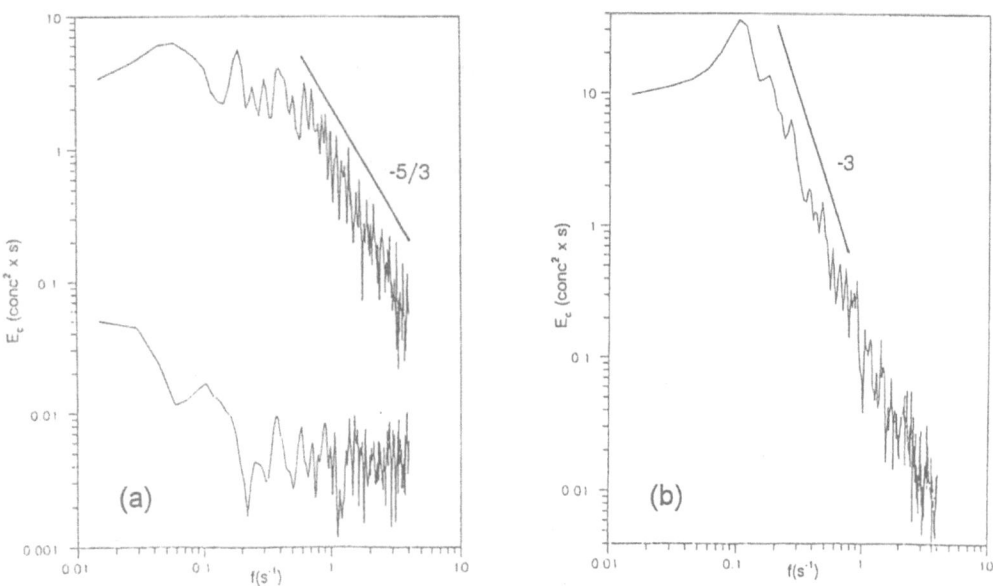

Fig. 20: Concentration fluctuation spectra for a shallow jet (B = 1 cm, H = 2.5 cm) obtained from LIF images for the two locations indicated in Fig. 15. a) Point 1 in the "middle field". The lower trace measures the background noise. b) Point 2 in the "far-field".

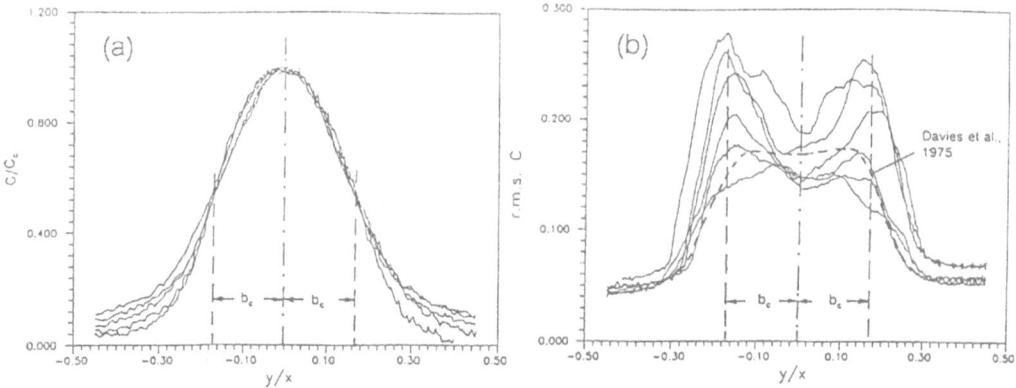

Fig. 21: Lateral profiles of a) mean concentration and b) rms concentration, both normalized by the centerline concentrations, for different locations in the shallow jet. b_c indicates the position of the concentration half-width.

3.5 Linear Stability Analysis of Shallow Flows

Much can be learned about the shallow jet behavior by means of a linear instability analysis of the two-dimensional depth-averaged equations of motion with turbulent bottom friction terms. The base flow is assumed as parallel. Obviously, given the gradual evolution in the jet far field, the results of such calculations are approximate at best. Nevertheless, they can explain qualitatively some of the observed jet features, as has been shown by earlier applications of this technique to other free shear flows (for example, see Huerre and Monkewitz, 1990, for the unbounded wake). In the following, the method is demonstrated for the shallow jet configuration in an ambient co-flow.

In a shallow water layer, the depth-averaged equations of motion for parallel, slightly disturbed flow with the two-dimensional (x,y) velocity field $(U + u, v)$, in which $U(y)$ is the base velocity and $u(x,y,t)$ and $v(x,y,t)$ are the disturbance velocities, are

$$\frac{\partial u}{\partial x} + \frac{\partial v}{\partial y} = 0 \tag{13}$$

$$\frac{\partial u}{\partial t} + U\frac{\partial u}{\partial x} + v\frac{\partial U}{\partial y} = -\frac{1}{\rho}\frac{\partial p}{\partial x} - \frac{c_f U}{H}u + \varepsilon_h \nabla^2 u \tag{14}$$

$$\frac{\partial v}{\partial t} + U\frac{\partial v}{\partial x} = -\frac{1}{\rho}\frac{\partial p}{\partial y} - \frac{c_f U}{H}v + \varepsilon_h \nabla^2 v \tag{15}$$

in which c_f is the turbulent friction coefficient and ε_h the horizontal eddy diffusivity. Equations of this type have first been suggested by Alavian and Chu (1985) for the shallow mixing layer. There are differences, however, in the appropriate choice of the horizontal diffusivity term for such flows (see below) and the computational technique.

Small amplitude disturbances for u and v are introduced that are harmonic in x,t, of the type $\phi e^{i(\alpha x - \beta t)}$ in which ϕ is a complex amplitude function. $\alpha = \alpha_r + \alpha_i$ where α_r is the wavenumber of the disturbance, and $-\alpha_i$ is the spatial amplification rate, and $\beta = \beta_r + i\beta_i$, where β_r is the frequency of the disturbance, and $-\beta_i$ is the temporal amplification rate. Substituting these terms into the depth-averaged equations of motions and eliminating the pressure and longitudinal velocity amplitude, one obtains the Orr-Sommerfeld equation with bottom friction

$$\left(U(y) - \frac{\beta}{\alpha} + \xi\right)\left(\phi'' - \alpha^2\phi\right) + \xi\frac{U'}{U}\phi' - U''\phi = \frac{\varepsilon_h}{i\alpha}\left(\phi'''' - 2\alpha^2\phi'' + \alpha^4\phi\right) \tag{16}$$

where the differentiation is with respect to y, $\xi = c_f U/(i\alpha H) = S_\ell U/(i\alpha 2\ell)$ is a parameter measuring the effect of local friction and $S_\ell = c_f 2\ell/H$ is a local stability parameter in which ℓ is the local jet half-width. ϕ is the eigenfunction which represents the amplitude of the disturbance in the y direction. The boundary conditions are

$$\phi(\pm L) = \phi'(\pm L) = 0 \tag{17}$$

which represents two side walls located at y = L and -L, where L/ℓ is sufficiently large (12.5 in the present calculations).

A self-preserving hyperbolic secant profile is assumed as a good approximation to the actual flow, as well as an exact solution of the Navier-Stokes equations,

$$\frac{U}{U_m - U_a} = R_j + \text{sech}^2 \left(\frac{y}{\ell}\right) \tag{18}$$

where $R_j = U_a/(U_m - U_a)$ is the jet velocity ratio, U_a the ambient co-flow velocity, U_m the jet centerline velocity, and ℓ the transverse length scale (ℓ is related to the half-velocity half-width b, b = $\sinh^{-1}(1) \ell = 0.881\ell$). $R_j = 0$ is a pure jet in an stagnant ambient, $R_j > 0$ a jet in ambient co-flow, and $R_j < 0$ a jet in ambient counter-flow.

Details on the actual computational procedure are given in Chen and Jirka (1998). The inviscid jet flow with its double inflection point on the velocity profile is necessarily unstable (Rayleigh condition) and there are two instability modes, the more unstable sinuous (anti-symmetric) and the less unstable varicose (symmetric) mode.

However, in the present case of a shallow turbulent jet there are two mechanisms that can cause stabilization:

1) The effect of lateral eddy viscosity is represented by the Reynolds number

$$Re_j = \frac{(U_m - U_a)\,\ell}{\varepsilon_h} \tag{19}$$

(in laminar jets the eddy viscosity would be replaced by the kinematic viscosity v), and

2) the effect of bottom friction acting over the jet width is represented by a shallow jet stability parameter

$$S = c_f \frac{2b}{H} \tag{20}$$

in analogy to the definition for the wake (Eq. 12).

A few salient results of the shallow jet stability calculations are provided here. First, Fig. 22 deals with a pure jet without any ambient co-flow ($R_j = 0$). Fig. 22a shows the marginal stability curves for the unbounded jet (S = 0) as function of Reynolds number and non-dimensional wavenumber $\alpha_r\ell$. The sinuous mode is the more unstable one starting at a mini-mum $Re_j = 5$, while the varicose mode starts at 90. In contrast, Fig. 22b displays the stability curves as a function of the shallowness parameter S for an "inviscid" jet ($Re_j \rightarrow \infty$), i.e. ne-glecting the role of the small scale eddy viscosity ($Re_j \rightarrow \infty$). Again, the sinuous mode is the more unstable one for the assumed sech^2 profiles. For this reason, in the remainder attention is paid to the sinuous mode alone.

The combined effect of viscosity and shallowness for a pure jet is shown in Fig. 23a as the critical shallowness parameter S_c as function of the inverse Reynolds number. Clearly, the effect of eddy viscosity is small whenever the Reynolds number is reasonably large, say $Re_j >$ 100. The critical shallowness parameter under these "inviscid" conditions is about 0.69. Finally, Fig. 23b demonstrates the effect of ambient co-flow, given by the parameter R_j, on a high Reynolds number shallow jet. Ambient co-flow has a significant stabilizing effect, while counter-flow destabilizes the shallow jet.

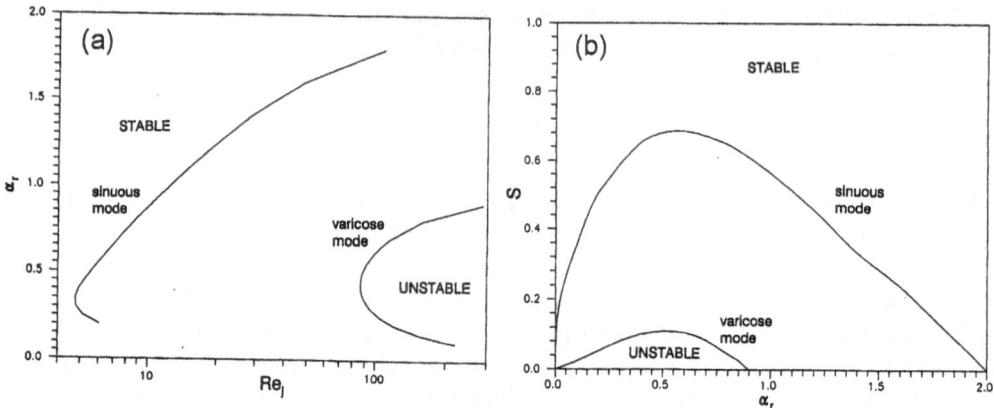

Fig. 22: Marginal stability curves for pure jet without ambient co-flow. a) The unbounded case (S = 0) showing wavenumber as a function of jet Reynolds number, based on eddy viscosity, Re_j and b) the shallow case with "inviscid" conditions ($Re_j \to \infty$) showing wavenumber as a function of the shallowness parameter S.

Comparison with observations: First, it can be shown that the shallow jet flows are controlled in the main by the turbulent bottom friction (expressed by the shallowness parameter S) rather than by lateral turbulent diffusive momentum exchange (expressed by the jet Reynolds number Re_j). A robust estimate for the transverse eddy viscosity in a wide open channel flow is $\varepsilon_h = 0.2\, u^* H$ (Fischer et al., 1979) where $u^* = \sqrt{c_f/2}\, U$ is the shear velocity and U is a depth-averaged velocity. Hence, the "viscous action" given by the local three-dimensional turbulence (length scale of order H) within the wide shallow jet flow is measured by choosing the jet velocity U_m as the velocity scale, so that ($U \approx U_m$ and $\ell \approx b$) the jet Reynolds number $Re_j \approx \left(7/\sqrt{c_f}\,(2\,b/H)\right)$. For the present experiments, with $c_f \le 0.01$ and $(2b)/H \approx$ (2 to 100), this evaluates to Re_j of the order of 200 to 10,000. The lower value applies to the start of the far field at which $b \approx H$. Yet larger values could hold for actual environmental conditions. Thus, the turbulent shallow jet indeed can be assumed as "inviscid" (Re_j) in the sense that small scale turbulence damping (with scales of the order of H) is negligible. The jet friction parameter S (or S_b) is the single controlling factor.

The range for the smooth wall friction factor for the experiments is c_f from 0.005 to 0.01. Hence, the high aspect ratio jets ($H/B \ll 1$), have at the beginning of the far field, where $b \approx H$, a typical S value from 0.01 to 0.02. Thus, referring to Fig. 23a, they are highly unstable and prone to the lateral amplification of any disturbance of which there are plenty in the highly turbulent flow emanating from the middle field. The disturbances grow quite rapidly as they are advected by the evolving jet flow. Their growth becomes quickly non-linear so the present linear theory is no longer valid to predict their evolution. Nevertheless, the onset of instability conforms well to the predictions of the linear theory. For example, the critical wavenumber at instability is $\alpha_r \ell \approx 0.6$ as a measure for the most amplified waves (Fig. 22b).

But $\alpha_r\ell$ is related to the Strouhal number of the dominant eddies as they are advected by a measurement point, $\alpha_r\ell = 2\pi St_p$. This predicts $St_p \approx 0.09$ in close agreement with the measured values.

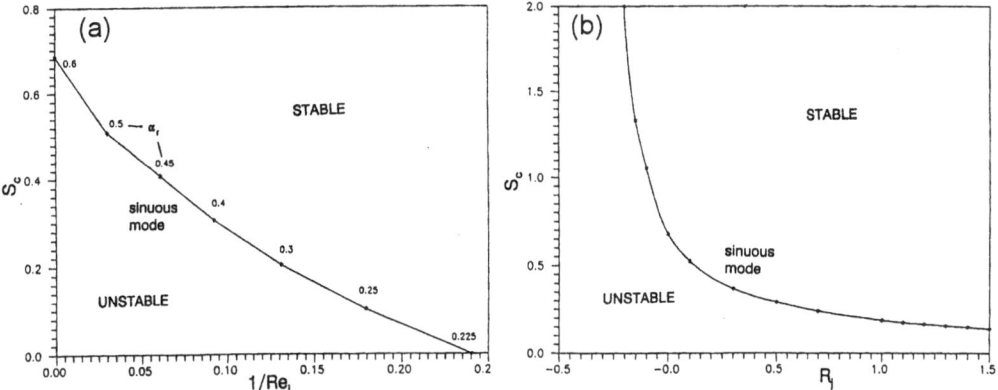

Fig. 23: Critical shallowness parameter S_c for the sinuous mode only. a) As a function of the inverse Reynolds number $1/Re_j$ for a pure jet without co-flow, and b) as a function of the jet co-flow parameter R_j for high Re_j "inviscid" conditions.

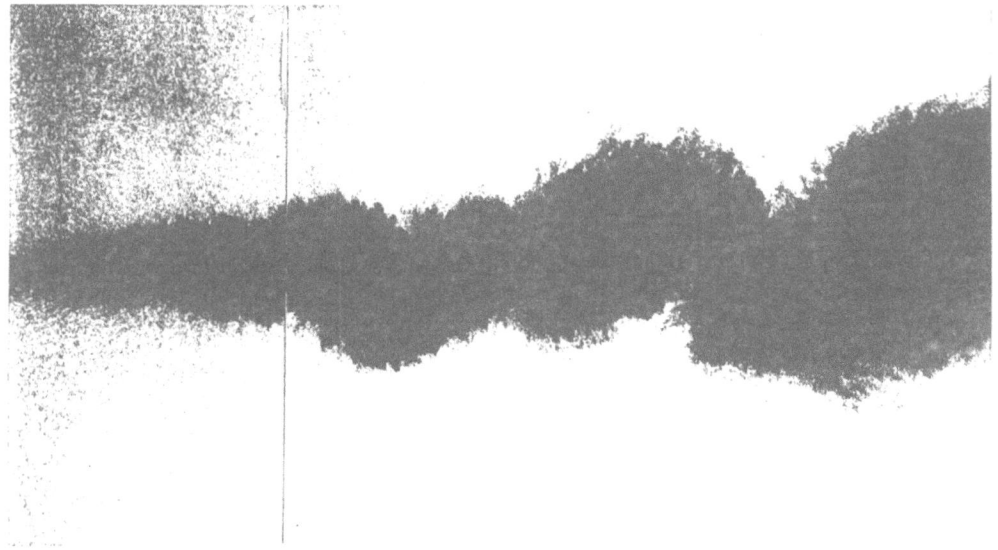

Fig. 24: Evolution of a shallow jet (H/B = 0.25, B = 10 cm, H = 2.5 cm, observed over about 2.5 m or 100 H). The flow is initially wide b/H > 1, but does not exhibit sinuous instabilities until much later. However, varicose modes are visible in the initial stage..

In principle, it appears possible to generate an initially stable shallow jet. This could be done by a low aspect ratio inflow, $H / B \gg 1$. In particular, when $B \approx 100\,H$ this should provide a stable jet, <u>assuming</u> the initial inflow is distributed according to the fully sheared profile, Eq.18. Such an initially wide jet could not be studied within the experimental set-up. However, Fig. 24 shows the evolution of a shallow jet with a low aspect ratio, $H / B = 0.25$. Indeed, the jet appears to be stable to sinuous deflections over a considerable initial distance. On the other hand, some varicose (symmetric) perturbations appear to be present over that initial distance. P. Huerre (private communication) has suggested that the full (top-hat) initial jet profile may actually be stable to sinuous perturbations, but unstable to varicose ones. Shallow jet stability analyses with alternate more full-bodied jet profiles (see Monkewitz, 1988) may be useful to resolve this question, but has not been performed to date.

4   Expert System CORMIX for the Prediction of Pollutant Releases into Water Bodies

The control of pollution sources from municipal or industrial point sources in order to achieve good water quality is one of the goals of sustainable water management. One of the principal approaches for that purpose is the enactment of water quality standards for different pollutant types (chemical and/or biological parameters). This procedure is being followed in water quality legislation both in the USA (USEPA, 1991) as well as the countries of the European Union (Commission, 1997; see also Ragas et al., 1997). The principal instrument of U.S. water quality regulations that has evolved since 1972 is the concept of the "mixing zone". This is a spatially defined and limited zone within a water body at the edge of which ambient water quality standards must be met while inside they can be exceeded. The U.S. Environmental Protection Agency (or its surrogate State water quality authorities) have given numerical definitions of such mixing zones and of ambient water quality standards for different types of water bodies. The definition of the mixing zone may also be given on ad hoc basis between the applicant seeking a discharge permit and the regulating authority: observations on existing discharges or computations for future discharges can be used to demonstrate that the mixing zone will be limited and the discharge will assure "the protection and propagation of a balanced, indigenous community of fish, shellfish and wildlife".

The use of mathematical predictive models has become an important tool in this permitting process. Such models are used by both the discharger (or consulting engineer) and the regulatory authority to predict the shape of the pollutant discharge plume in order to define associated mixing zones. Furthermore, the determination of detailed features, such as degree of contact with the bottom or shoreline of the water body, often is required by biologists in order to assess impact on ecologically sensitive habitats or key species.

The three major discharge modes for waste water discharges are the submerged single port, the submerged multiport diffuser, and the surface discharge. The latter type - by means of a simple open channel inflow - is, of course, the oldest and most economical type of discharge structure. Higher mixing rates can be achieved with the submerged discharge modes, however.

Numerous mathematical predictive models mostly of the jet-integral type and some higher-order turbulence numerical models, have been developed in recent years. *Jet-integral models* consist of a set of ordinary differential equations derived from the cross-sectional (normal to the jet trajectory) integration of jet-properties such as mass, momentum, and buoyancy fluxes. Empirical formulations for internal jet behavior such as buoyant damping of turbulence and cross-sectional distortion (lateral spreading) are included. The equation systems are parabolic and are solved by simple forward-marching numerical schemes along the jet trajectory. Jet-integral models perform satisfactorily for simple flows with no shoreline interaction or attachment. However, strong crosscurrents or limited depths causing attachment with the downstream bank or strong initial buoyancy causing intrusion of the effluent along the upstream bank render these models invalid. In addition, jet-integral models predict only the jet-like behavior of the flow near the source. They are incapable of simulating any far-field processes that occur after a certain transition distance (Jirka et al., 1981).

Three-dimensional *numerical models with higher-order turbulence closure* schemes attempt to approximate the system of Reynolds equations through finite element or finite difference schemes. To a large extent, these methods have not been successful for practical routine applications. The major problem lies in the specification of boundary conditions. In addition, the formulation of turbulent transport terms under the influence of buoyancy effects and with different resolution detail in the near- and far-field poses obstacles. From a practical viewpoint the models are highly complicated, difficult to check, and expensive to use, even in moderate size test cases.

Thus, to date, there is no generally valid and universally applicable predictive model for surface or submerged discharges with their multiple complexities of jet mixing, boundary interaction, buoyant collapse and convection processes, ambient water depths ranging from very shallow to deep, and lateral limitations imposed by the shorelines. In the absence of such a general model, the recent development of CORMIX - the Cornell Mixing Zone Expert System - pursued a different two-step methodology. The system first provides a classification of the plume type that pertains to the user-specified combination of ambient and discharge conditions. A dynamic length-scale approach is consistently used for that purpose as explained in more detail in the following. In the second predictive step, CORMIX uses a series of partly analytical and partly numerical modules to calculate the detailed plume features for the given flow class. CORMIX has found wide application in engineering practice and many cases of successful comparison with detailed laboratory data as well as field data have been reported.

CORMIX (Doneker and Jirka, 1991; Jirka et al. 1996) consists of three subsystems that can be accessed in an PC-environment: CORMIX 1 for the analysis of submerged single port discharges, CORMIX 2 for submerged multiport diffusers and CORMIX 3 for buoyant surface discharges. In the following, some of the modeling philosophy and technical approach for the surface discharge option is explained in more detail.

As all presently available predictive models for discharge analysis, CORMIX is a steady-state model. Thus, the user has to select appropriate design conditions that approximate the often time-variable conditions in the environment. In many instances, this does not pose a major problem as discharge-induced mixing processes occur reasonably fast relative to the scale of natural flow variability, e.g. in river flows or lake currents. A major, heretoforeto unresolved, problem exists, however, in the highly transient plume motions in estuarine and coastal flows during tidal reversal. The plume shape is observed to rapidly change and re-entrainment of previously discharged mixed effluent water can occur in the velocity reversal immediately after slack tide. Consequently, steady-state models become spatially highly limited and will tend to underpredict actual maximum concentrations (temperature rises) in the reversal process. Detailed experimental observations and similarity relationships are reported below as a first step toward an improved methodology for including such effects in predictive models, such as CORMIX.

4.1 Steady-State Behavior of Surface Plumes

Earlier studies have considered the mixing of buoyant surface jets under two limiting conditions. First, discharges into a stagnant ambient (e.g. Hayashi and Shuto, 1968) have shown an initial strongly entraining jet region followed by the buoyant damping and collapse leading to an unsteady buoyant pool with horizontally spreading density fronts. Second, discharges into a uniform steady crossflow (e.g. Abdelwahed and Chu, 1968) exhibit time-invariant behavior with gradual deflection of the surface plume and final advection by the crossflow. Appropriate length scales for these processes and predictive models (usually in form of jet integral equations) have been developed and shown to be reasonably accurate measures of the mixing processes under these limiting conditions (see Jirka et al.,1981; Chu and Jirka, 1986; Jones et al., 1996).

The structure of a buoyant surface discharge and its interaction with a steady receiving environment is described primarily by the discharge buoyancy flux $J_o = g_o' U_o a_o$, momentum flux $M_o = U_o^2 a_o$, the discharge flow $Q_o = U_o a_o$, and the receiving water body's depth H and velocity u_a (see Fig.25), in which U_o is the discharge velocity, $a_o = b_o h_o$ the cross-sectional area with channel width b_o and channel depth h_o, and g_o' the discharge buoyant acceleration arising from the discharge temperature excess above the ambient temperature. Dynamic length scales can be formed from these four parameters using dimensional analysis: First, the transition where the initial jet momentum becomes dominated by buoyant spreading is described by the *jet-to-plume length scale* L_M

$$L_M = \frac{M_o^{3/4}}{J_o^{1/2}} \tag{21}$$

For a distance less than L_M momentum dominates the flow and therefore jet mixing prevails, for a distance greater than L_M buoyancy dominates and strong lateral spreading prevails. As shown in Fig.25, the scale L_M is also related to the maximum buoyant surface jet depth h_{max} whose magnitude relative to the ambient depth H controls in turn whether a given discharge situation behaves as deep or shallow. Second, the *jet-to-crossflow length* L_m scale measures the relative significance of the initial momentum and the ambient crossflow velocity

$$L_m = \frac{M_o^{1/2}}{u_a} \tag{22}$$

L_m is a measure of the position (see Fig.25) where the flow changes from the weakly deflected regime to the strongly deflected regime. Third, the *plume-to-crossflow length scale* L_b measures the relative importance of the initial buoyancy flux to the ambient crossflow velocity

$$L_b = \frac{J_o}{u_a^3} \tag{23}$$

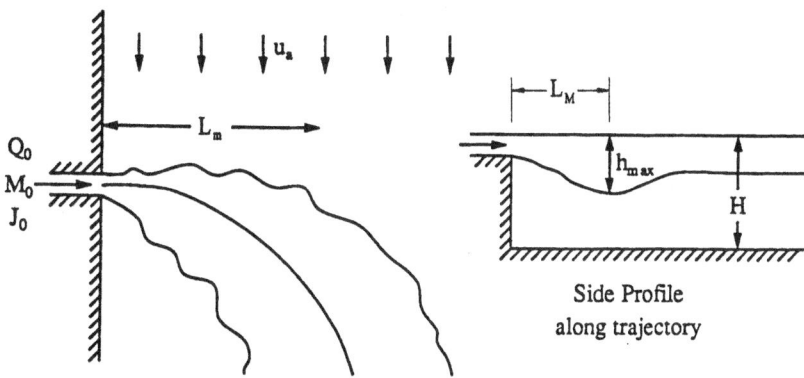

Fig. 25: Plan and cross-sectional views of buoyant surface discharge in steady crossflow.

L_b has a significantly different meaning for surface plumes than for submerged buoyant jets. Since this length scale represents an interaction of the initial buoyancy of the effluent and the velocity of the crossflow, its most apparent measure is the extent of upstream spreading that a surface plume may exhibit. It also plays a role in the increased lateral progression of free jets caused by the thinning of a buoyant surface jet due to buoyancy. Fourth, the *discharge length scale* L_Q measures the significance of the volume flux as compared to the momentum flux

$$L_Q = \frac{Q_o}{M_o^{1/2}} \tag{24}$$

This length scale is essentially a size measure for the discharge cross-section and generally of lesser importance. It plays a critical role, however, when measured against L_M in determining whether upstream intrusion occurs or not. Finally, note that some of the scales given above are interrelated, such as $L_M = L_m^{3/2} / L_b^{1/2}$; nevertheless, all of the scales are useful indicators in their own of separate flow processes.

Comparing these four length scales with each other and with the ambient depth H, respectively, yields a classification system for buoyant surface discharges. Fig.26, taken from Chu and Jirka (1986) represents laboratory and field data observations for different surface jet and plume types, all with a cross-flowing (i.e. pointing offshore) discharge. More recently, this has been expanded by Jones et al. (1995) for an arbitrary discharge orientation and additional dependence on the channel aspect ratio $A = h_o/b_o$. Fig.27 shows the CORMIX classification system for the four major surface discharge classes: 1) **Free jets** that are not attached to the shoreline with varying degrees of vertical mixing over the available water depth, 2) **Shoreline-attached jets** with near-shore recirculation zones, 3) **Wall jets** whenever the discharge angle is small so that the jet centerline attaches to and then follows the shoreline, and 4) **Surface plumes** that are governed by weak jet momentum (low discharge densimetric Froude numbers, i.e. L_M/L_Q) leading to either strong upstream spreading in weak crossflow or rapid deflection in strong crossflow. All criterion values (C1, C3 to C7) are shown in Fig.27 as being of order of unity. Their actual numerical value can not, of course, be obtained from dimensional analysis and must be determined from applicable data (see Jones et al., 1996, for detail).

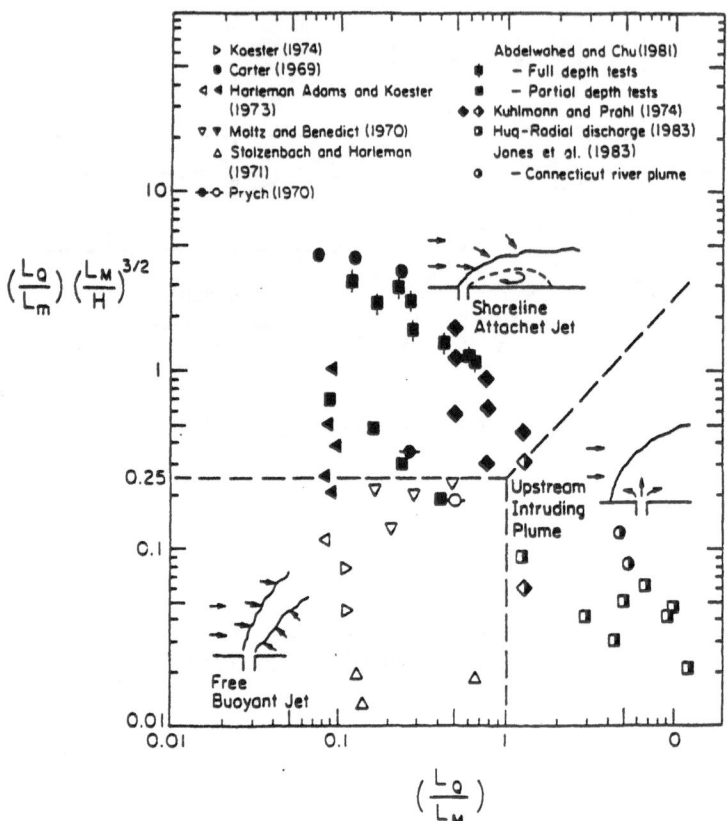

Fig. 26: Laboratory and field data indicating different surface buoyant jet types (from Chu and Jirka, 1985).

Numerous observations from field or laboratory confirm the above classification system. As an example, Fig.28 shows several laboratory observations in which a Planar Laser Induced Fluorescence (PLIF) technique (Nash et al., 1995) was used to measure temperature levels in the heated discharge flow. The four observations represent short-time observations (with about 10 sec averaging time) of a a) free jet class FJ1, b) a shoreline-attached jet class SA2, c) a wall jet class WJ2, and d) an upstream intruding plume class PL1, respectively. The structural differences among these flow types are considerable as seen in these plan views. Other details, such the extent of vertical mixing and bottom contact (as it exists for the SA2 and WJ2 examples) must be inferred from other measurements, not shown here.

Clearly, different types of predictive models must be employed to deal with this demonstrated variety of discharge plume conditions. In some instances, the jet/plume can be assumed as free so that simple jet-integral models are applicable. In other cases of strong shoreline or bottom contact or of strong upstream intrusion that is clearly not valid and other approaches must be used. In the CORMIX predictive system a number of different flow modules are employed in their appropriate spatial sequence to provide a full prediction from the near- to far-field for all these possibilities (see Jones et al., 1996). This has been tested with numerous comparisons to detailed laboratory or field data. For example, Fig. 29 shows a comparison with field data for cooling water discharges from two power plants located on Lake Michigan, USA.

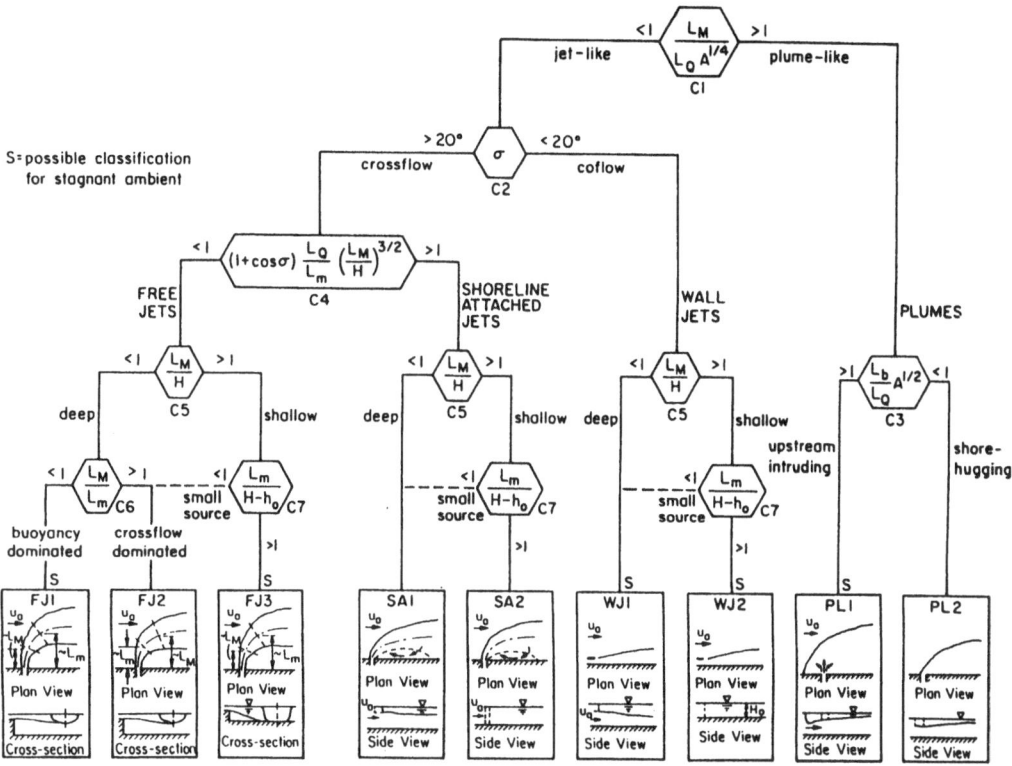

Fig. 27: CORMIX3 classification diagram for buoyant surface discharges (Jones et al., 1996).

4.2 Unsteady Behavior of Surface Plumes in Tidal Reversing Currents

Ambient flow conditions in natural water bodies are practically always unsteady. In some environments, such as rivers, lakes or reservoirs, the unsteadiness is usually moderate so that steady-state design conditions can be readily specified for the application of mathematical predictive models.

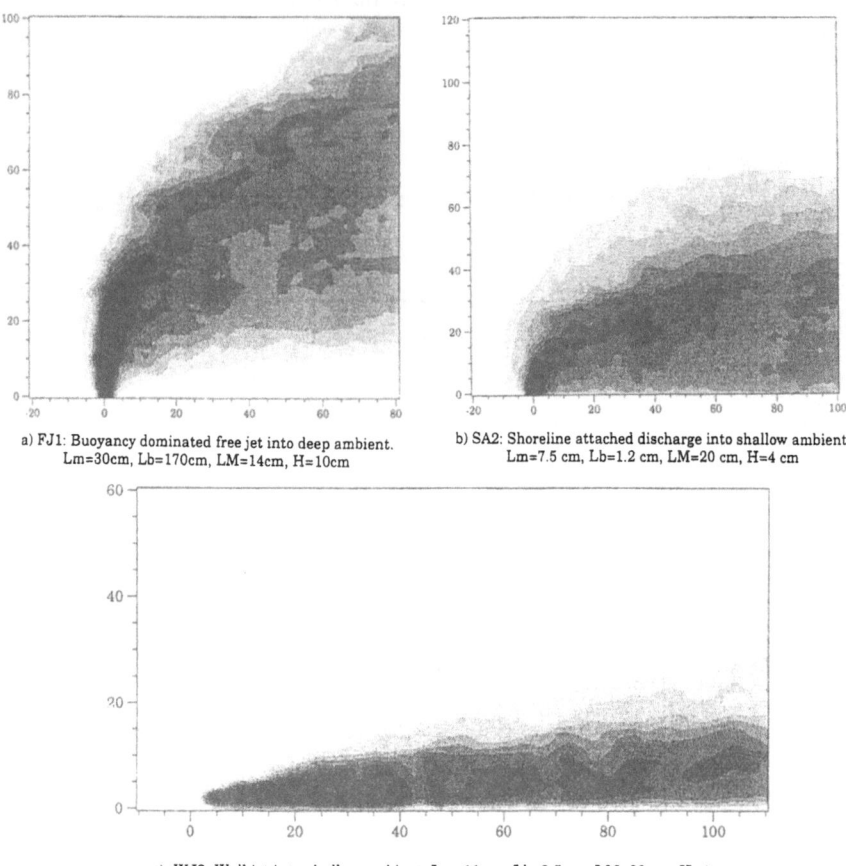

a) FJ1: Buoyancy dominated free jet into deep ambient.
Lm=30cm, Lb=170cm, LM=14cm, H=10cm

b) SA2: Shoreline attached discharge into shallow ambient.
Lm=7.5 cm, Lb=1.2 cm, LM=20 cm, H=4 cm

c) WJ2: Wall jet into shallow ambient: Lm=11 cm, Lb=3.5 cm, LM=20 cm, H=4 cm

Fig. 28: Plan views of surface buoyant jets obtained by planar laser-induced fluorescence (PLIF) mapping (from Jones et al, 1996): a) Free jet class FJ1, b) shoreline-attached jet class SA2, c) wall jet class WJ2, and d) upstream intruding plume class PL1.

Estuarine or coastal flow conditions, however, are determined by the interplay of tidal variations with a 12.4 h time period, freshwater inflows, and wind driven currents. Strongly unsteady conditions with current reversals at slack tide are the hallmark of these flows. Two major effects on plume behavior are to be expected under these conditions: First, a reduced plume offshore extent over which the plume is able to adjust in a near-steady fashion to the instantaneous ambient flow conditions. Outside this extent the plume will be distorted and lag behind. Second, a build-up of concentration (temperature rise) in the phase after reversal as re-entrainment into the near-field zone of previously mixed water occurs.

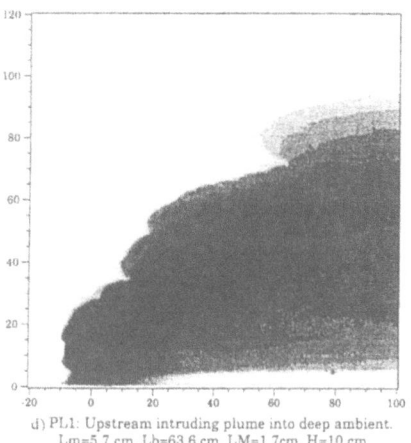

d) PL1: Upstream intruding plume into deep ambient.
Lm=5.7 cm, Lb=63.6 cm, LM=1.7cm, H=10 cm

Fig. 28: Continued

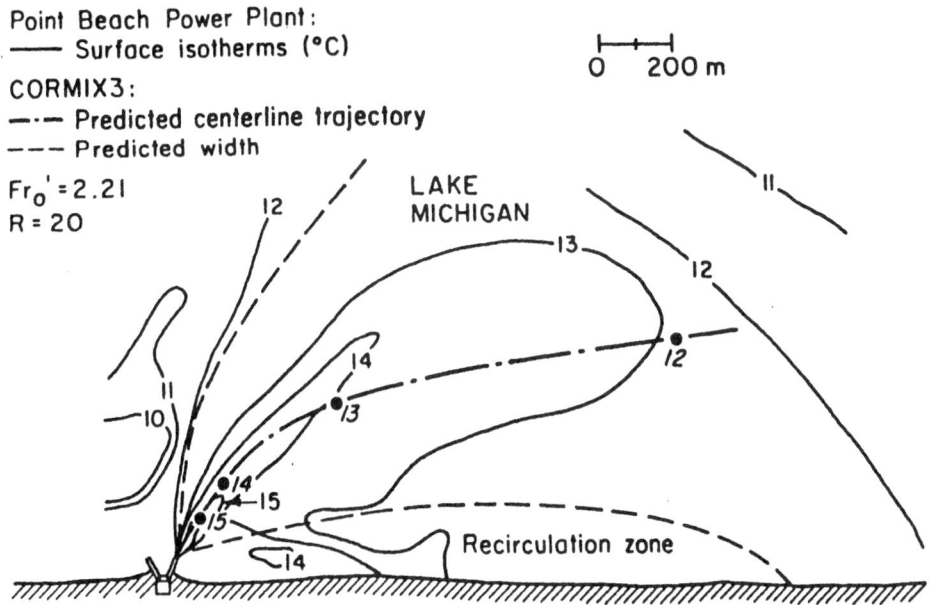

Fig. 29: Surface isotherms of the cooling water discharge from thermal-electric power plants located at Lake Michigan and corresponding CORMIX3 predictions. a) Point Beach Power Plant, b) Palisades Nuclear Power Plant (from Jones et al., 1996).

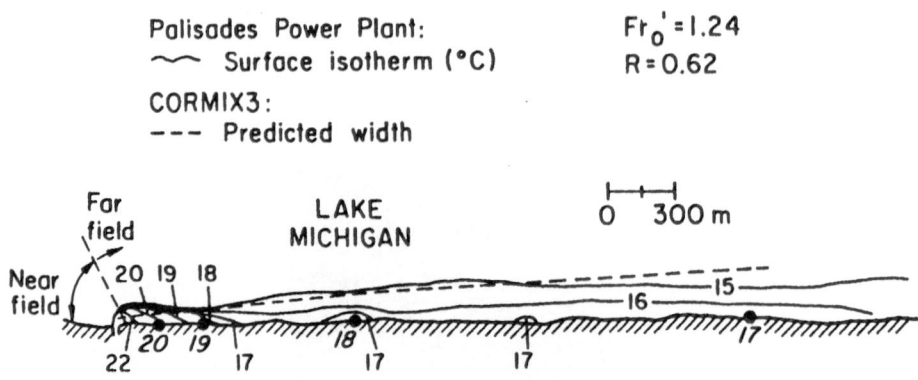

Palisades Power Plant:

~ Surface isotherm (°C)

$Fr_0' = 1.24$

$R = 0.62$

CORMIX3:

- - - Predicted width

Fig. 29: Continued

Fig.30 illustrates the large variation in ambient tidal velocity profiles $u_a(t)$ that may arise from specific coastal conditions. This variation gives rise to different rates of acceleration $du_a(t)/dt$, a parameter found to be critical in characterizing unsteady flows. The variation of the jet-to-crossflow length scale L_m over a sinusoidal tidal half-cycle is displayed in Fig.31. The physical dimensions given in the figure correspond to the conditions of a 200 MW power plant ($Q_o = 10$ m^3/s, $U_o = 10$ m/s, $M_o = 10$ m^4/s^2, $\Delta T_o = 20$ °C, $g'_o = 0.04$ m/s^2, $J_o = 10$ m^4/s^3). Near slack tide, $u_a(t) \rightarrow 0$, L_m becomes unbounded and thus is an unsatisfactory measure of the buoyant jet behavior under these transient low-velocity conditions.

A preferred measure for describing the unsteady trajectory and the build-up of the buoyant pool in this transient low-velocity phase is given by a relationship between discharge momentum flux M_o and the ambient acceleration $du_a(t)/dt$, which, in contrast to L_m, remains finite as $u_a(t) \rightarrow 0$. On dimensional grounds, this leads to the *jet-to-unsteady-crossflow length and time scales* (reversal scales) as defined by Nash and Jirka (1996)

$$L_u = \left(M_o \bigg/ \left[\frac{du_a}{dt} \right] \right)^{1/3} \qquad T_u = \left(M_o \bigg/ \left[\frac{du_a}{dt} \right]^4 \right)^{1/6} \qquad (25)$$

Although other scales can be formed from the interaction between the discharge buoyancy flux J_o and ambient acceleration, $du_a(t)/dt$, those scales are not considered dominant, following Jirka et al. who showed that buoyant surface jet deflection in crossflow is primarily influenced by the discharge momentum, not the buoyancy. Physically, the reversal length scale is representative of the distance at which the effects of acceleration become appreciable. Fig.31 shows the interplay between the two length scales L_m and L_u for a typical coastal discharge. During most of the tidal cycle, $L_m < L_u$, so that the ambient acceleration is negligible compared to the instantaneous velocity. However, as slack tide is approached, $L_m > L_u$, and the reversal length scale becomes the dominant influence.

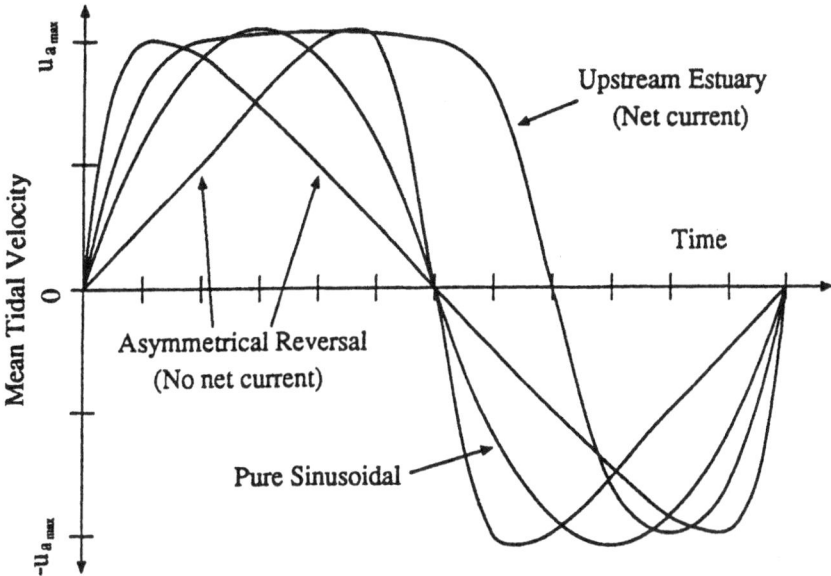

Fig. 30: Schematic examples of tidal cycle velocity variations showing different rates of ambient acceleration at slack tide.

The amount of the pollutant build-up resulting from buoyant accumulation of discharge can be related to the time during which the ambient can be considered quiescent (slack tide). If the ambient velocity is approximated as having a linear variation with time i.e. $u_a(t) = [du_a(t)/dt]_o\, t$ then the duration of slack tide, as defined by the intersection of L_u and L_m in Fig. 31, is given by $t = \pm T_u$. The time scale T_u can thus be interpreted as the "duration of slack tide". For most flows, this time scale is much greater than the intrinsic discharge time scale $T_M = M_o / J_o$, indicating that there is a relatively long time period during which the instantaneous velocity is unimportant, when the plume may fully evolve through the influence of the unsteadiness alone (represented by L_u). Thus, for unsteady flows, the nondimensional parameter of L_M / L_u is proposed here to replace L_M / L_m to uniquely characterize the geometry and mixing of the flow during reversal. In addition, this ratio also describes the duration of slack tide, as $L_M / L_u = (T_M / T_u)^{1/2}$, and thus reflects the temporal as well as the spatial behavior of the discharge.

An approximate 1:100 scale model, based on densimetric Froude number similarity, was chosen to investigate the near field behavior of the prototype thermal surface discharge specified above with an ambient depth of $H = 10$ m. A continuous time series of surface temperature mappings was obtained in a 6 m x 8 m x 20 cm deep reversing flow basin in the

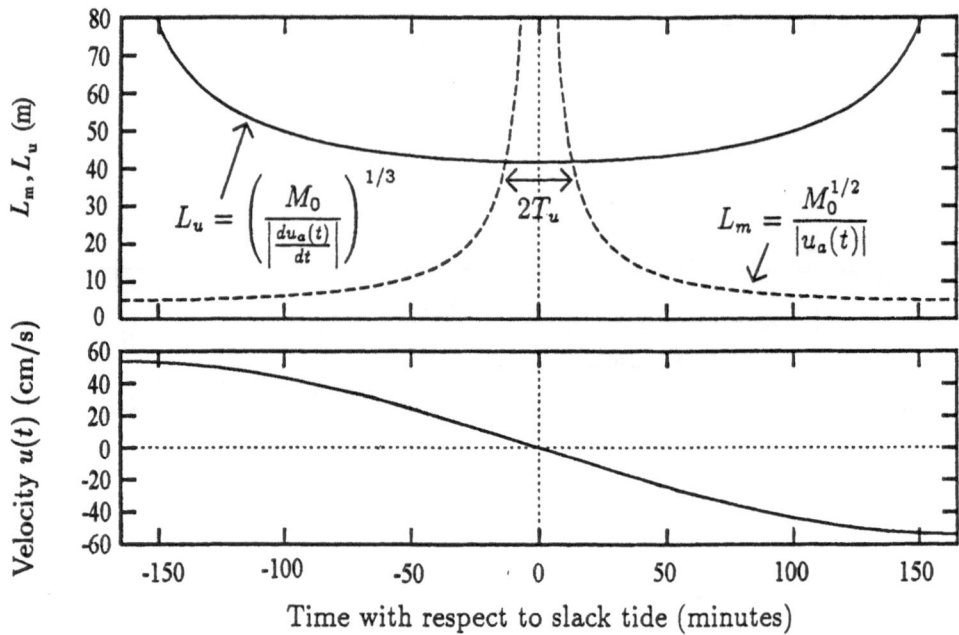

Fig. 31: Variation of the length scales L_m and L_u during a sinusoidal change in ambient velocity $u_a(t)$.

DeFrees Hydraulics Laboratory using planar laser-induced fluorescence (PLIF). This basin is large enough to contain the entire plume evolution while the effects of unsteadiness are significant ($-T_u < t < T_u$), which is the important consideration in this experiment. All of the experimental results displayed below are given in scaled prototype dimensions.

Four linearly transient velocity variations with constant $du_a(t)/dt$, ranging from very rapid (0.02 cm/s²) to very mild (0.005 cm/s²), and varying between velocity plateaus of $u_{amax} = +60$ cm/s were chosen to represent typical environmental flows. This linear representation is a good approximation in the time frame of reversal (see Fig.30). For comparison, Fig.32 shows an acceleration around reversal of 0.0085 cm/s². A typical time series of near-surface temperature distributions, as indicated by the grey scales of the video images obtained from the PLIF method, is shown in Fig.32 for a case of a surface jet in rapidly reversing flow (L_u / L_M = 2.4). The mappings reveal the asymmetrical behavior of the jet before and after reversal, and the maximum induced temperature rise occurring slightly after slack tide. The effects of unsteadiness become negligible outside the duration of slack tide ($t < T_u$ and $t > T_u$), and mappings having the same instantaneous velocity become symmetric before and after reversal.

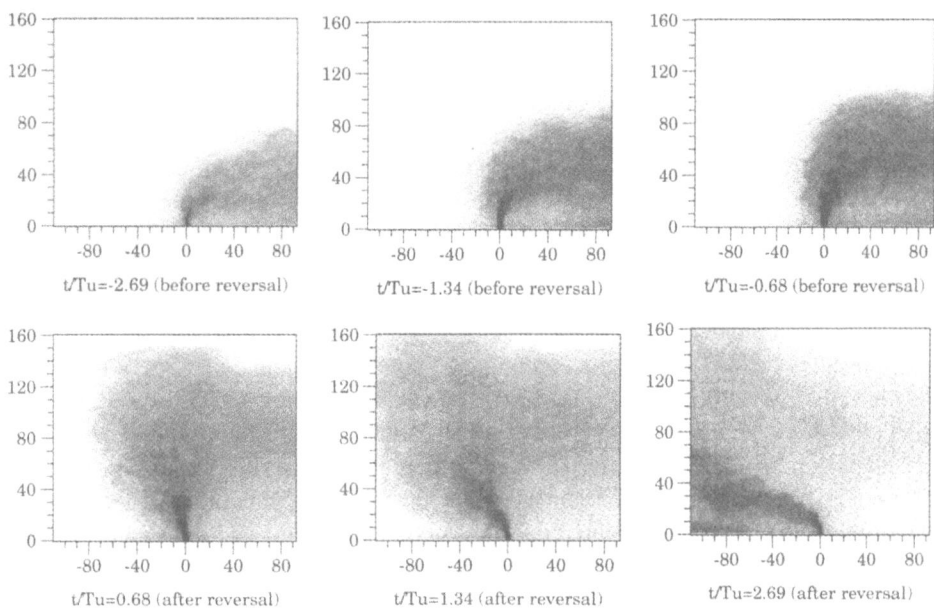

Fig. 32: Time series of instantaneous surface temperature mappings (plan view) by means of PLIF technique for a surface buoyant jet in tidal reversing current. Distances are displayed in m. Images are shown for different non-dimensional times t/T_u.

The detailed analysis of all the unsteady simulations conducted shows that the time evolution of plume trajectory and mixing has, indeed, a substantial dependence on the rate of ambient acceleration, $du_a(t)/dt$, given in relative terms by the scales L_u and T_u. The experiments reveal, as hypothesized, that the temperature buildup, or pool formation around slack tide is more severe in less rapid reversals, as there is a longer duration for buoyant accumulation. In contrast, discharges into quickly reversing currents have little time for near field accumulation, but have highly asymmetrical trajectories before and after slack tide.

A buoyant surface discharge can be described by its centerline trajectory and buoyancy (temperature) decay along that trajectory. Some unique, approximately self-similar, properties of the unsteady jet behavior can be extracted from the complete data series if the appropriate scale measures, L_u and T_u, are employed. This has been demonstrated for two unsteady plume measures: the maximum centerline temperature rise at a given distance along the trajectory, and the centerline trajectory (Nash and Jirka, 1996).

In summary, surface buoyant discharges represent complicated flow and mixing phenomena if the full spectrum of ambient conditions ranging from deep to shallow water depths, stagnant to weak to strong crossflows, lateral confinement, and variable discharge fluxes and buoyancy are to be considered. Yet a consistent classification of all possible flow phenomena is possible by means of the several governing dynamic length scales that control the specific discharge-ambient interaction. This classification is part of the Cornell Mixing Zone Expert System (CORMIX). Together with its predictive modules this length scale classification enables CORMIX to calculate the detailed plume features for a wide variety of steady-state flow conditions.

It has been shown that buoyant surface discharges in unsteady ambient environments also exhibit self-similar geometry and mixing characteristics. However, correct time and length scales must be used to elicit these properties. The unsteady reversal conditions, in particular, are governed by a relation between the discharge momentum flux and the ambient flow acceleration, giving rise to the jet-to-unsteady-crossflow length and time scales, L_u and T_u. In particular, T_u can be seen as a consistent dynamic measure of the "duration of slack tide". Plume properties such as their limited offshore extent and their concentration (temperature rise) build-up effects in this highly unsteady phase can be uniquely represented when scaled appropriately by L_u and T_u.

The results obtained from these experimental simulations and from others (e.g. Brocard, 1985) have been incorporated into CORMIX as modifications to its steady state predictions; principally, these modifications consist of a truncation of the spatial plume extent and of a build-up factor that describes the pollutant re-entrainment in this highly unsteady phase. Traditionally, the analysis of slack, or near-slack low velocity, conditions has been the main focus of regulatory mixing zone analysis since mixing is weakest and induced concentrations are highest in this stage. Thus, a more realistic assessment of these critical design conditions appears possible.

Acknowledgements:

The author's research in these areas of environmental fluid mechanics has been supported at Cornell University by U.S. National Science Foundation, U.S. Environmental Protection Agency, Electric Power Research Institute, National Oceanic and Atmospheric Administration, and U.S. Geological Survey, and at the University of Karlsruhe by the Deutsche Forschungsgemeinschaft.

5   References

Abdelwahed, M.S.T., and Chu, V.H. (1968). Surface Jets in Plumes and Crossflows. *Technical Report*, 81-1 (FML), McGill University, Montreal.

Alavian, V., and Chu, V.H. (1985). Turbulent exchange in shallow compound channel. In *Proceedings 21 Congress International Association of Hydraulic Research*, Melbourne, Australia.

Asher, W. E., and Pankow, J. F. (1986). The interaction of mechanically generated turbulence and interfacial films with a liquid phase controlled gas/liquid transport process. *Tellus*, 38B, 305-318.

Batchelor, G.K. (1969). Computation of the energy spectra in homogenous two-dimensional turbulence. *Physics of Fluids*, 233-238.

Benilov, A.Y., Kouznetsov, O.A. , and Panin, G.N. (1974). On the analysis of wind wave-induced disturbances in the atmospheric turbulent surface layer. *Boundary Layer Meteorology*, 6, 269-285.

Brocard, D.N. (1985). Surface buoyant jets in steady and reversing crossflows. *Journal of Hydraulic Engineering*, 111(5).

Broecker, Ch. (1982). The influence of bubbles upon gas exchange. In *NATO Advanced Study Institute on Air-Sea Exchange of Gases and Particles*, Durham, New Hampshire.

Broecker, W.S., and Peng, T.-H. (1982). *Tracers in the Sea*, Eldigio Press.

Brumley, B.H. and Jirka, G.H. (1987). Near-surface turbulence in a grid-stirred tank. *Journal of Fluid Mechanics*, 183, 235-263.

Brumley, B.H., and Jirka, G.H. (1988). Air-water transfer of slightly soluble gases: turbulence, interfacial processes and conceptual models. *Journal of Physico-Chemical Hydrodynamics*, 10, 3, 1988.

Brutsaert, W., and Jirka, G.H., Ed.s, (1984). *Gas Transfer at Water Surfaces*, Reidel Publishing Company.

Chen, D. and Jirka, G.H. (1991). Pollutant mixing in wake flows behind islands in shallow water. In *Proceedings Int. Symp. on Environmental Hydraulics* (J.H.W. Lee and Y.K. Cheung, Ed.s), Balkema, 371-377.

Chen, D. and Jirka, G.H. (1995). Experimental study of plane turbulent wake in a shallow water layer. *Fluid Dynamics Research*, 16, 11.

Chen, D. and Jirka, G.H. (1997). Absolute and convective instabilities of plane turbulent wakes in a shallow water layer. *Journal of Fluid Mechanics*, 338, 157-172.

Chen, D. and Jirka, G.H. (1998). Linear instability analyses of turbulent mixing layers and jets in shallow water layers. *Journal of Hydraulic Research*, 36, No.5, 815-830.

Chen, D. and Jirka, G.H. (1999). A laser-induced fluorescence study of a plane shallow jet. *Journal of Hydraulic Engineering* (in press).

Chu, C.-R. (1993). Experiments on gas transfer and turbulence structure in free surface flows with combined wind/bottom shear. *Ph.D. Thesis, Cornell University*, Ithaca, New York

Chu, C.-R., and Jirka, G.H. (1992). Turbulent gas flux measurements below the air-water interface of a grid-stirred tank. *International Journal of Heat and Mass Transfer*, 35 (8), 1957-68.

Chu, V.H. and Babarutsi, S. (1988). Confinement and bed-friction effects in shallow turbulent mixing layers. *Jounrnal of Hydraulic Engineering*, 114, 1257-1274.

Chu, V.H., and Jirka, G.H. (1986). Buoyant surface jets and plumes in environmental fluid mechanics. Chapter 27 in *Encyclopedia of Fluid Mechanics*. N. Cheremisinoff (Ed.), Gulf Publishing Co.

Chu, V.H., Wu, J.H. and Khayat, R.E. (1983). Stability of turbulent shear flows in shallow channel. In *Proceedings XX Congress IAHR*, Moscow, 3, 128-133.

Coantic, M. (1980). Mass transfer across the ocean-air interface: small scale hydrodynamic and aerodynamic mechanisms. *Journal of Physico-Chemical Hydrodynamics*, 1, 249-279.

Commission of the European Communities (1997). Proposal for a counsel directive establishing a framwork for Community action in the field of water policy. COM(97) 47 final, Brussels.

Davies, A.E., Keffer, J.F. and Baines, W.D. (1975). Spread of a heated plane turbulent jet. *Physics of Fluids*, 18, 770.

Deacon, E.L. (1977). Gas transfer to and across an air-water interface. In *Tellus*, 29, 363-374.

Doneker, R.L., and Jirka. G.H. (1991). Expert Systems for Design and Mixing Zone Analysis of Aqueous Pollutant Discharges. *Journal of Water Resources Planning and Management*, 117, No. 6, 679-697.

Dracos, T., Giger, M. and Jirka, G.H. (1992). Plane Turbulent Jets in a Bounded Fluid Layer. *Journal of Fluid Mechanics*, 214, 587-614.

Fischer, H.B., List, E.J., Koh, R.C.Y., Imberger, J., and Brooks, N.H. (1979). *Mixing in Inland and Coastal Waters*. Academic Press, New York.

Fortescue, G.E., and Pearson, J.R. (1967). On gas absorption into a turbulent liquid. *Chemical Engineering Science*, 22, 187-216.

Giger, M., Dracos, T. and Jirka, G.H. (1991). Entrainment and mixing in plane turbulent jets in shallow water. *Journal of Hydraulics Research*, 29, No.4, 615-643.

Hardy, J.T. (1982). The sea surface microlayer: Biology, chemistry and anthropogenic enrichment. *Progress in Oceanography*, 11, 307-328.

Hayashi, T., and Shuto, N. (1967). Diffusion of warm water jets discharged horizontally at water surface. In *Proceedings 12th Congress of the International Association of Hydraulic Research*, Fort Collins, Colorado.

Hoover, T.E., and Berkshire, D.C. (1969). Effects of hydration on carbon dioxide exchange across an air-water interface. *Journal of Geophysical Research*, 74, 456-464.

Hopfinger, E. J., and Toly, J. A. (1976). Spatially decaying turbulence and its relation to mixing across density interfaces. *Journal of Fluid Mechanics*, 78, 155-175.

Huerre, P., and Monkewitz, P.A. (1990). Local and global instabilities in spatially developing flows. *Annual Review of Fluid Mechanics*, 22, 473-537.

Hunt, J. C. R. (1984). Turbulence structure and turbulent diffusion gas-liquid interface, *Gas Transfer at Water Surfaces*, Reidel Publishing Company.

Hunt, J. C. R., and Graham, J. M. R. (1978). Free-stream turbulence near plane boundaries. *Journal of Fluid Mechanics*, 84, 209-235.

Hussain, A.K.M.F. (1983). Coherent structures - reality and myth. *Physics of Fluids*, 26, 2816-2850.

Ingram, R.G., and V.H. Chu (1987). Flow around islands in Rupert Bay: An investigation of the bottom friction effect. *Journal of Geophysical Research*, 92(C13), 14521-14533.

Jahne, B., et al. (1987). On the parameters influencing air-water gas exchange. *Journal of Geophysical Research*, 92, C2, 1937-49.

Jahne, B., et. al. (1984) Wind/wave-tunnel experiment on the Schmidt number and wave field dependence of air/water gas exchange. In *Gas Transfer at Water Surfaces*, Reidel Publishing Company.

Jähne, B., and Haußecker, H. (1998). Air-water gas exchange. *Annual Review of Fluid Mechanics*, 30, 443-68.

Jirka, G.H. (1994). Shallow Jets. In: *Recent Advances in the Fluid Mechanics of Turbulent Jets and Plumes*, P.A. Davies and M.J. Valente Neves (Ed.s), Kluwer Academic Publishers, Dordrecht.

Jirka, G.H., Adams, V, and Stolzenbach, K.D. (1981). Properties of buoyant surface jets. *Journal of the Hydraulics Division*, ASCE, 107, HY 11.

Jirka, G.H., and Brutsaert, W. (1984). Measurements of wind effects in water-side controlled gas exchange in natural rivers. In *Gas Transfer at Water Surfaces*, Reidel Publishing Company.

Jirka, G.H., and Ho, A. W.-K. (1990). Gas transfer at the water surface: Measurements of gas concentration fluctuations. *Journal of Hydraulic Engineering*, 116, 6, 835-847.

Jirka, G.H., Doneker, R.L., and Hinton, S.W. (1996). User's Manual for CORMIX: A Hydrodynamic Mixing Zone Model and Decision Support System for Pollutant Discharges into Surface Waters. *Technical Report, DeFrees hydraulics Laboratory, Cornell University* (also published b y U.S. Environmental Protection Agency. Technical Report, Environmental Research Lab. Athens, Georgia).

Jones, G.R., Nash, J.D., and Jirka, G.H. (1996). CORMIX3: An Expert System for the Analysis and Prediction of Buoyant Surface Discharges. *Technical Report, DeFrees Hydraulics Laboratory, School of Civil and Environmental Engineering, Cornell University* (also published by U.S. Environmental Protection Agency, Tech. Rep., Environmental Research Lab, Athens, Georgia).

Kemp, M., and Boynton, W. (1980). Influence of biological and physical processes on dissolved oxygen dynamics in an estuarine system. *Estuarine, Coastal and Marine Science*, 1, 407-431.

Kitaigorodskii, S.A. (1984). On the fluid dynamical theory of turbulent gas transfer across an air-sea interface in the presence of breaking wind-waves. *Journal of Physical Oceanography*, 4, 960-972.

Kraichnan, R. (1967). Inertial ranges in two-dimensional turbulence. *Physics of Fluids*, 10, 1417-1428.

Lewis, W.K., and Whitman, W.G. (1924). Principles of gas absorption. *Industrial and Engineering Chemistry*, 16, 1215-1220.

Lion, L.W. (1984). The surface of the ocean. In *Handbook of Environmental Chemistry*, 1, Part C., O. Hutzinger, Ed., Springer.

Liss, P.S. (1973). Processes of gas exchange across an air-water interface. *Deep-Sea Research*, 20, 221-238.

Liss, P.S., and Slater, P.G. (1974). Flux of gases across the air-sea interface. *Nature*, 247, 181-184.

Liss, P.S., Balls, P.W., Martinelli, F.N., and Coantic, M. (1981). The effect of evaporation and condensation on gas transfer across and air-water interface. *Oceanological Acta*, 4, 129-138.

Lloyd, P.M. and Stansby, P.K. (1997a). Shallow-water flow around model conical islands of small side slope. I: Surface piercing. *Journal of Hydraulic Engineering*, 123, No. 12, 1057-1067.

Lloyd, P.M. and Stansby, P.K. (1997b). Shallow-water flow around model conical islands of small side slope. II: Submerged. *Journal of Hydraulic Engineering*, 123, No. 12, 1068-1077.

MacDonald, D.G. and Jirka, G.H. (1997). Characteristics of headland wakes in shallow flow. In *Proceedings XXVII Congress IAHR*, San Francisco, Vol.1, 88-93.

Mackay, D., and Yuen, A.T.K. (1983). Mass transfer coefficient correlations for volatilization of organic solutes from water. *Environmental Science and Technology*, 17, 211-217.

Marino, R., and Howarth, R.W. (1992). Atmospheric oxygen exchange in the Hudson River: dome measurements and comparison with other natural waters. *Estuaries*, 17.

Melville, W.K., Rapp R.J., and Chan, E.-S. (1985). Wave breaking, turbulence and mixing. In *The Ocean Surface: Wave Breaking, Turbulent Mixing and Radio Probing*, Y. Toba and H.Mitsuyasu, Ed.s, Reidel Publishing Company.

Merlivat, L., and Memery, L. (1983). Gas exchange across an air-water interface: experimental results and modeling of bubble contribution to transfer. *Journal of Geophysical Research*, 88C, 707-724.

Moog, D.B. (1995). Experiments on open channel gas transfer with large-scale roughness elements. *Ph.D. thesis, Cornell University*, Ithaca, New York.

Moog, D.B., and Jirka, G.H. (1999a). Air-water gas transfer in uniform channel flow. *Journal of Hydraulic Engineering*, 125, 1, 3-10.

Moog, D.B., and Jirka, G.H. (1999b). Stream reaeration in non-uniform channel flow: Macro roughness enhancement. *Journal of Hydraulic Engineering*, 125, 1, 11-16.

Nash, J.D., and Jirka, G.H. (1996). Buoyant surface discharges in unsteady ambient flows. *Dynamics of Atmosphere and Oceans*, 24, 75-84

Nash, J.D., Jirka, G.H., and Chen, D. (1995). Large-scale planar laser-induced fluorescence measurements in turbulent density-stratified flows. *Experiments in Fluids*, 19, 297-304.

Nezu, I. und Nakagawa, H. (1993). *Turbulence in Open-Channel Flows*. A.A. Balkema, Rotterdam.

O'Connor, D.J. (1983). Wind effects on gas-liquid transfer coefficients. *Journal of Environmental Engineering*, 109, 731-752.

Quinn, J.A., and Otto, N.C. (1971). Carbon dioxide exchange at the air-sea interface: Flux augmentation by chemical reaction. *Journal of Geophysical Research*, 76, 1539-49.

Ragas, A.M.J., Hams, J.L.M., and Leuv n, R.S.E.W. (1997). Selecting water quality models for discharge permitting. *European Water Pollution Control*, 7(5), 59-67.

Rathbun, R.E. (1988). Discussion of "Flume tests on hydrocarbon reaeration tracer gases" by J.D. Bales and E.R. Holley. *Journal of Environmental Engineering*, 114, 2.

Su, M.Y., Green, A.W., and Bergin, M.T. (1984). Experimental studies of surface wave breaking and air entrainment. In *Gas Transfer at Water Surfaces*, Reidel Publishing Company.

Thames Survey Committee and Water Pollution Research Laboratory (1964). Effects of polluting discharges on the Thames estuary. London, 349-363.

Theofanous, T.G. (1984). Conceptual models of gas exchange. In *Gas Transfer at Water Surfaces*, Reidel Publishing Company.

Thomas, F. O. and Goldschmidt, V.W. (1986). Structural characteristics of developing turbulent planar jet. *Journal of Fluid Mechanics*, 63, 227-256.

Uijttewaal, W.S.J. and Tukker, J. (1998). Development of quasi two-dimensional structures in a shallow free-surface mixing layer. *Experiments in Fluids*, 24, 192-200.

U.S. Environmental Protection Agency (1981). *Technical Support Document for Water Quality-based Toxics Control*. Office of Water, Washington, DC, Report No. EPA 505/2-90-001.

Van Heijst, G.J., Clerx, H. and Maassen, S. (1996). Stably stratified flow in a rectangular container: cell pattern formation and anomalous diffusion. In *5th IMA Conference on Stratified Flows*, Dundee, Scotland.

Wilhelms, S., and Gulliver, J., Ed.s, (1991). *Proceedings of Second International Symposium on Gas Transfer at Water Surfaces*. American Society of Civil Engineers, New York.

Wu, J. (1988). Bubbles in the near-surface ocean: A general description. *Journal of Geophysical Research*, 93, CL, 587-590.

COMPUTATIONAL ENVIRONMENTAL GEOMECHANICS

B.A. Schrefler
University of Padua, Padua, Italy

Abstract. This paper presents a general framework for the computational analysis of environmental geomechanics problems. It is based on heat and multiphase flow in deforming porous media where pollutant transport mechanisms can be added. The governing equations are derived and then discretised by means of the finite element method in space and finite differences in time. Appropriate solution methods are addressed. Examples given involve heat and mass transfer together with pollutant transport in deforming geomaterials and surface subsidence problems.

1 Introduction

Environmental geomechanics is a rather new topic in the field of mechanics. Only in recent years have some scientific conferences dealing with this topic been held. Environmental problems in geomechanics are essentially multi-physics problems and often involve the transport of some substance. Transport of contaminants and other substances may occur in underground fluids, e.g. in water, water vapour and air, filling the pores of geomaterials. Mass transport also takes place in reservoir engineering problems, where the fluids involved are oil, water and gas. Transport phenomena alone have been well studied but much less so their effects in connection with the deformation of the solid matrix. This connection is peculiar to environmental geomechanics.

Typical problems of the area of environmental geomechanics are now listed. Some deal with soil pollution due to chemical pollutants, heavy metals and non aqueous phase liquids (see Gambolati and Verri 1995, Schrefler et al. 1994). Pollutant transport in aquifers has received much attention while its transport in partially saturated zones is a rather recent topic of research. This zone is important however because contaminants enter groundwater through it. Contaminants may also be released into the atmosphere through the vadose zone. This is the case if we have volatilisation of contaminants such as chlorinated solvents and gasoline hydrocarbons from shallow groundwater and the migration of contaminants to the soil surface through diffusion. Coupling of mechanical effects with transport phenomena is important in the vadose zone because of the nature of the soil and capillary effects.

Another issue of concern is the confinement of industrial, municipal and nuclear waste, where the design of appropriate barriers is important (Olivella et al. 1992, Manassero and Shakelford 1994). Common materials for such barriers are soil, rock or concrete. The behaviour of these materials in deep and surface disposal conditions has to be studied from the point of view of safety

[1] I wish to thank my co-workers G. Bolzon, D. Gawin, C.E. Majorana, L. Sanavia, L. Simoni, X. Zhan, who over years have contributed to the work reported in this paper. This work has been partly financed by research funds M.U.R.S.T. 60%.

and durability. Finally, an important topic with environmental implications is surface subsidence due to either groundwater withdrawal or as experienced above exploited hydrocarbon reservoirs. Capillary effects at reservoir level seem to play a role in the latter case as far as they may produce either swelling of the formation if the reservoir is in the elastic range or irreversible volumetric compaction if the yield locus is reached (Bolzon et al. 1996, Schrefler et al. 1997).

This paper presents a common theoretical framework on which to model the above mentioned problems and numerical solutions based on the finite element method (Schrefler 1995, Lewis and Schrefler 1998). The porous medium (geomaterials) is assumed to be a multiphase system where interstitial voids of the solid matrix may be filled with water, water vapour, dry air (gas) and contaminants. These may or may not be miscible with the fluid phases present in the pores, but immiscible and generally non reacting with the solid phase. To handle this multiphase system correctly, the general frame of averaging theories is used in deriving the governing equations.

The ensuing numerical model solves the linear momentum balance equation for the whole multiphase medium, mass balance equations for aqueous and non-aqueous fluid phases, the energy (enthalpy) balance equation for the multiphase medium and the mass transport equation for pollutant in the aqueous and non-aqueous phases. In the case of immiscible contaminant, the last two equations are substituted by the mass balance for that substance.

The numerical examples will show applications of the full set of equations and its subsets. In particular two pollutant transport examples and subsidence above a compacting gas reservoir will be investigated. Other examples can be found in (Schrefler 1995).

2   The Mathematical Model

2.1   Kinematic Equations

The multiphase medium will be described as the superposition of all π phases, $\pi = 1, 2,...\kappa$, i.e. in the actual configuration each spatial point $\mathbf{x}$ is simultaneously occupied by material points $\mathbf{X}^\pi$ of all phases. The state of motion of each phase is however described independently. Based on these assumptions, the kinematics of a multiphase medium is dealt with next.

In a Lagrangian or material description of motion, the position of each material point $\mathbf{x}^\pi$ at time t is function of its placement in a chosen reference configuration, $\mathbf{X}^\pi$ and of the current time t

$$\mathbf{x}^\pi = \mathbf{x}^\pi\left(\mathbf{X}^\pi, t\right) \qquad (1)$$

To keep this mapping continuous and bijective at all times, the Jacobian $\mathbf{J}$ of this transformation must not equal zero and must be strictly positive, since it is equal to the determinant of the deformation gradient tensor $\mathbf{F}^\pi$

$$\mathbf{F}^\pi = \text{Grad } \mathbf{x}^\pi \qquad \left(\mathbf{F}^\pi\right)^{-1} = \text{grad } \mathbf{X}^\pi \qquad (2)$$

Because of the non-singularity of the Lagrangian relationship (1), its inverse can be written and the Eulerian or spatial description of motion follows

$$\mathbf{X}^\pi = \mathbf{X}^\pi\left(\mathbf{x}^\pi, t\right) \qquad (3)$$

The material time derivative of any differentiable function $f^\pi(\mathbf{x}, t)$ given in its spatial description and referred to a moving particle of the π phase is

$$\frac{\overset{\pi}{D} f^\pi}{Dt} = \frac{\partial f^\pi}{\partial t} + \operatorname{grad} f^\pi \mathbf{v}^\pi \tag{4}$$

If superscript α is used for $\dfrac{D}{Dt}$, the time derivative is taken moving with the α phase.

The deformation process of the solid skeleton is described by the velocity gradient tensor, $\mathbf{L}^s$ which, referred to spatial co-ordinates, is given by

$$\mathbf{L}^s \equiv \operatorname{grad} \mathbf{v}^s = \mathbf{D}^s + \mathbf{W}^s \tag{5}$$

Its symmetric part, $\mathbf{D}^s$, is the Eulerian strain rate tensor, while its skew-symmetric component, $\mathbf{W}^s$, is the spin tensor.

2.2 Microscopic Balance Equations

The microscopic situation of any π phase is described by the classical equations of continuum mechanics. At the interfaces to other constituents, the material properties and thermodynamic quantities may present step discontinuities.

For a thermodynamic property Ψ the conservation equation within the π phase may be written as

$$\frac{\partial(\rho\Psi)}{\partial t} + \operatorname{div}(\rho\Psi\dot{\mathbf{r}}) - \operatorname{div} \mathbf{i} - \rho\mathbf{b} = \rho\mathbf{G} \tag{6}$$

where $\dot{\mathbf{r}}$ is the local value of the velocity field of the π phase in a fixed point in space, $\mathbf{i}$ is the flux vector associated with Ψ, $\mathbf{b}$ the external supply of Ψ and $\mathbf{G}$ is the net production of Ψ. The relevant thermodynamic properties Ψ are mass, momentum, energy and entropy. The values assumed by $\mathbf{i}$, $\mathbf{b}$ and $\mathbf{G}$ are given in (Hassanizadeh and Gray 1979a, Lewis and Schrefler 1998). The constituents are assumed to be microscopically non polar, hence the angular momentum balance equation is here omitted. This equation shows however that the stress tensor is symmetric.

2.3 Macroscopic balance equations

In this section the macroscopic balance equations for mass, linear momentum, angular momentum and energy (enthalpy) are obtained and then specialised for a deforming porous material, where heat transfer, flow of water (liquid and vapour), dry air and miscible or immiscible pollutants (e.g. dense non-aqueous phase liquids DNAPL) is taking place. For a proper description of the noniso-thermal unsaturated porous medium we need to take into account not only heat conduction and vapour diffusion, but also heat convection, fluid flow due to pressure gradients or capillary effects and latent heat transfer due to water phase change (evaporation and condensation) inside the pores.

Furthermore the solid is deformable, resulting in coupling of the fluid, solid and thermal fields. All fluid phases are in contact with the solid phase.

We must also consider pollutants. Immiscible pollutants behave as a fluid phase alone while for soluble pollutants we have three transport processes: advection, diffusion and dispersion. Mass flux of contaminants due to advection is linked to pressure gradients, diffusive mass flux is due to concentration gradients and dispersive flux (mechanical dispersion) is attributed to variations of seepage velocity during migration.

The constituents are assumed to be chemically non reacting. A local thermodynamic equilibrium assumption will be used so that the temperatures of each constituent at a point in the multiphase medium are taken to be equal. This does not mean that the temperature is uniform throughout the medium but only that at each point one temperature is sufficient to characterise the state. Momentum exchanges due to mechanical interaction are independent of temperature gradient. The effect of solid deformation in the energy balance equation is neglected. The stress is defined as tension positive for the solid phase, while pore pressure is defined as compressive positive for fluids.

As far as pollutants are concerned the following additional hypotheses are made: mechanical properties of the medium are not changed by the pollutant; no effects of miscible pollutants are accounted for in the linear and angular momentum balance and the energy balance equation.

The macroscopic balance equations are obtained by systematically applying averaging principles as developed by Hassanizadeh and Gray (1979a-b, 1980) to the microscopic balance equation (6), where for each constituent the generic thermodynamic variable ψ is replaced by appropriate microscopic properties. For more details see Lewis and Schrefler (1998). In this process the volume fractions η^π appear which are identified as follows

solid phase $\eta^s = 1 - n$ where $n = \dfrac{dv^w + dv^g}{dv}$ is porosity and dv^π is the volume of constituent π within a representative elementary volume R.E.V.;

water $\eta^w = n S_w$ where $S_w = \dfrac{dv^w}{dv^w + dv^g}$ is the degree of water saturation and

gas $\eta^g = n S^g$ with $S_g = \dfrac{dv^g}{dv^w + dv^g}$ the degree of gas saturation.

It follows immediately that

$$S_w + S_g = 1 \tag{7}$$

It is reminded that gas is a mixture of dry air and vapour.

Taking into account the kinematics of section 2.1 we obtain the mass balance equation for the solid as

$$\frac{\overset{s}{D}(1-n)\rho^s}{Dt} + \rho^s(1-n)\operatorname{div}\overline{v}^s = 0 \tag{8}$$

where

$$\rho^{\pi} = \frac{1}{dv^{\pi}} \int_{dv} \rho \gamma^{\pi} dv_m \qquad (9)$$

is the intrinsic phase averaged density and γ^{π} the distribution function; $\overline{v}^{\pi}$ is mass averaged velocity (Lewis and Schrefler 1998).

The macroscopic mass balance equation for water is

$$\frac{\overset{w}{D}}{Dt}\left(n S_w \rho^w\right) + n S_w \rho^w \text{div } \overline{v}^w = -\dot{m} \qquad (10)$$

where $-\dot{m}$ is the quantity of water per unit time, lost through evaporation (mass rate of water evaporation).

The macroscopic mass balance equation for gas, which is a mixture of dry air (ga) and vapour, (gw) may be written as

$$\frac{\overset{g}{D}\left(n S_g \rho^g\right)}{Dt} + n S_g \rho^g \text{ div } \overline{v}^g = \dot{m} \qquad (11)$$

where

$$\rho^g = \rho^{ga} + \rho^{gw} \qquad (12)$$

is the density of the mixture,

$$\overline{v}^g = \frac{1}{\rho^g}\left(\rho^{ga} \overline{v}^{ga} + \rho^{gw} \overline{v}^{gw}\right) = c^{ga} \overline{v}^{ga} + c^{gw} \overline{v}^{gw} \qquad (13)$$

with

$$c^{\pi} = \rho^{\pi} / \rho^g \qquad (14)$$

the mass fraction of component π, subject to $\sum_{\pi} c^{\pi} = 1$, $\pi = gw, ga$.

The mass balance equation for water vapour is

$$\frac{\overset{g}{D}}{Dt}\left(n S_g \rho^{gw}\right) + \text{div } \mathbf{J}_g^{gw} + n S_g \rho^{gw} \text{div } \overline{v}^g = \dot{m} \qquad (15)$$

where

$$\mathbf{J}_g^{gw} = n\, S_g \rho^{gw} \mathbf{u}^{gw} \tag{16}$$

is the diffusive-dispersive mass flux of component gw and

$$\mathbf{u}^\pi = \overline{\mathbf{v}}^{\pi g} = \overline{\mathbf{v}}^\pi - \overline{\mathbf{v}}^g \tag{17}$$

the macroscopic diffusive-dispersive velocity, $\pi = ga, gw$, subject to

$$\rho^{ga}\mathbf{u}^{ga} + \rho^{gw}\mathbf{u}^{gw} = \rho^g \sum_\pi c^\pi \mathbf{u}^\pi = 0 \tag{18}$$

The macroscopic mass balance equation for dry air is not needed because we use eq. (11) instead.

The mass balance equation for an immiscible pollutant π is the same as (10) with $\dot{m} = 0$.

$$\frac{D^\pi}{Dt}\left(n\, S_\pi \rho^\pi\right) + n\, S_\pi \rho^\pi \mathrm{div}\, \overline{\mathbf{v}}^\pi = 0 \tag{19}$$

The mass balance equations for soluble pollutants, propagating in the fluid phases, are similar to eq. (15). These equations are however usually written in terms of concentrations (14), where the index w or g has been added to specify if the pollutant is soluble in water or in the gas phase.

For concentration in water c_w^π we therefore have

$$\frac{D^w}{Dt}\left(n S_w c_w^\pi\right) + \mathrm{div}\, \mathbf{J}_w^\pi + n S_w c_w^\pi \mathrm{div}\, \overline{\mathbf{v}}^w = I_w \tag{20}$$

and for concentration in the gaseous phase c_g^π

$$\frac{D^g}{Dt}\left(n S_g c_g^\pi\right) + \mathrm{div}\, \mathbf{J}_g^\pi + n S_g c_g^\pi \mathrm{div}\, \overline{\mathbf{v}}^g = I_g \tag{21}$$

where I_π is interphase change and $\mathbf{J}_g^\pi$ has a similar meaning to eq. (16).

The macroscopic linear momentum balance equation for the solid phase becomes

$$\mathrm{div}\, \mathbf{t}^s + \rho_s\left(\overline{\mathbf{g}}^s - \overline{\mathbf{a}}^s\right) + \rho_s\, \hat{\mathbf{t}}^s = \mathbf{0} \tag{22}$$

where $\mathbf{t}^\pi$ is the stress tensor also called partial stress tensor, ρ_π the phase averaged density (equal to intrinsic phase averaged density ρ^π multiplied by volume fraction η^π), $\overline{\mathbf{g}}^\pi$ is external momentum supply, related to gravitational effects, $\overline{\mathbf{a}}^{\pi}$ acceleration, $\hat{\mathbf{t}}^{\pi}$ accounts for exchange of momentum due to mechanical interaction with other phases and sub/superscript $\pi = s$.

The macroscopic linear momentum balance for fluid phases is

$$\text{div } \mathbf{t}^{\pi} + \rho_{\pi}\left(\overline{\mathbf{g}}^{\pi} - \overline{\mathbf{a}}^{\pi}\right) + \rho_{\pi}\left[e^{\pi}(\rho\,\dot{\mathbf{r}}) + \hat{\mathbf{t}}^{\pi}\right] = 0 \tag{23}$$

where as well as the quantities defined above $\rho_{\pi}e^{\pi}(\rho\,\dot{\mathbf{r}})$ appears, the sum of the momentum exchange due to averaged mass supply and the intrinsic momentum supply due to a change of density and referred to the deviation $\tilde{\mathbf{r}}$ of the velocity of constituent π from its mass averaged velocity $\overline{\mathbf{v}}^{\pi}$. This last term indicated as $e^{\pi}(\rho\,\tilde{\mathbf{r}})$ will also appear in eq. (26). The average linear momentum balance equations are subject to the constraint

$$\sum_{\pi}\rho^{\pi}\left[e^{\pi}(\rho\,\dot{\mathbf{r}}) + \hat{\mathbf{t}}^{\pi}\right] = 0. \tag{24}$$

The average angular momentum balance equation shows that for non-polar media the partial stress tensor is symmetric $\mathbf{t}^{\pi} = \left(\mathbf{t}^{\pi}\right)^{\mathrm{T}}$ at macroscopic level also and the sum of the coupling vectors of angular momentum between the phases vanishes (Hassanizadeh and Gray 1979b).

The macroscopic energy balance equation is

$$\rho_{\pi}\frac{D\overline{E}^{\pi}}{Dt} = \mathbf{t}^{\pi} \cdot \mathbf{L}^{\pi} + \rho_{\pi}h^{\pi} - \text{div } \mathbf{q}^{\pi} + \rho_{\pi}\left[e^{\pi}(\rho\hat{E}) - e^{\pi}(\rho)\overline{E}^{\pi}\right] + \rho_{\pi}Q^{\pi} \tag{25}$$

where $\overline{E}^{\pi}$ accounts for averaged specific energy and for averaged kinetic energy related to $\tilde{\mathbf{r}}$, h^{π} is the sum of averaged heat sources, $\mathbf{q}^{\pi}$ a macroscopic heat flux vector, $\rho_{\pi}e^{\pi}(\rho\hat{E})$ the exchange term of internal energy due to phase change and possible mass exchange between constituents, $\rho_{\pi}e^{\pi}(\rho)$ is quantity of π phase per unit time and volume, lost through phase change ($= -\dot{m}$ for water) and finally Q^{π} represents exchange of energy due to mechanical interaction.

The energy balance equations are subject to

$$\sum_{\pi}\rho_{\pi}\left[e^{\pi}(\rho\hat{E}) + e^{\pi}(\rho\,\tilde{\mathbf{r}})\cdot\overline{\mathbf{v}}^{\pi} + \frac{1}{2}e^{\pi}(\rho)\overline{\mathbf{v}}^{\pi}\cdot\overline{\mathbf{v}}^{\pi} + \hat{\mathbf{t}}^{\pi}\cdot\overline{\mathbf{v}}^{\pi} + Q^{\pi}\right] = 0 \tag{26}$$

Phase change and the corresponding supply terms will be considered in the following for fluid phases only. Through averaging procedures the entropy inequality for the mixture may also be obtained as (Hassanizadeh and Gray 1979b, Lewis and Schrefler 1998)

$$\sum_{\pi}\left[\rho_{\pi}\frac{D\overline{\lambda}^{\pi}}{Dt} + \rho_{\pi}e^{\pi}(\rho)\overline{\lambda}^{\pi} + \text{div}\left(\frac{1}{\theta^{\pi}}\mathbf{q}^{\pi}\right) - \frac{1}{\theta^{\pi}}\rho_{\pi}h^{\pi}\right] \geq 0 \tag{27}$$

where $\bar{\lambda}^\pi$ is the averaged entropy of constituent π, θ^π the absolute temperature (here considered equal in a point for all constituents) and $\dfrac{h^\pi}{\theta}$ the averaged entropy source.

3 Constitutive Equations

The constitutive equations are obtained from the entropy inequality written for the multiphase material, under assumption of equilibrium or through linearisation, following Hassanizadeh and Gray (1990) and Gray and Hassanizadeh (1991a-b).

The effects of microstructural interaction between solid and fluid phases are described globally through experimentally obtained constitutive parameters or relations, such as the capillary pressure saturation relationship, the relative permeability, etc. For the closure of the model we need some state equations and for the fluid-vapour interaction we assume the validity of the Kelvin-Laplace equation, which was derived in (Baggio et al. 1997). The relations used in the following are now briefly listed.

The stress tensor in the fluid phases is

$$t^\pi = -\eta^\pi p^\pi I \tag{28}$$

where I is the unit tensor and p^π the macroscopic pressure of π phase. It can immediately be seen that the stress vector in the fluid phase does not have any dissipating part. The effects of deviatoric stress components will be accounted for in linear momentum balance equations through momentum exchange terms.

The moist air (gas) in the pore system is assumed to be a perfect mixture of two ideal gases, dry air and water vapour. The equation of state of perfect gas is hence valid

$$\begin{aligned} p^{ga} &= \rho^{ga}\theta R / M_a \\ p^{gw} &= \rho^{gw}\theta R / M_w \end{aligned} \tag{29}$$

where M_π is the molar mass of constituent π and R the universal gas constant.

Further Dalton's law applies, giving the gas pressure p^g and the molar mass of the gas M_g as

$$p^g = p^{ga} + p^{gw}$$

$$M_g = \left(\frac{\rho^{gw}}{\rho^g} \frac{1}{M_w} + \frac{\rho^{ga}}{\rho^g} \frac{1}{M_a} \right)^{-1} \tag{30}$$

where ρ^g is defined by (12).

Water is usually present in the pores as a condensed liquid, separated from its vapour by a concave meniscus because of surface tension. The capillary pressure p^c is defined as

$$p^c = p^g - p^w \tag{31}$$

where p^w is pressure of liquid water. A similar relationship applies also for DNAPL, i.e. capillary pressure must be defined for each couple of fluid phases present in the multiphase medium.

For liquids the equation of state is

$$\frac{1}{\rho^\pi}\frac{D\rho^\pi}{Dt} = \frac{1}{K_\pi}\frac{Dp^\pi}{Dt} - \beta_\pi\frac{DT}{Dt} - \gamma_\pi\frac{Dc^\pi}{Dt} \tag{32}$$

where T is the temperature above a reference value θ_0, such that $T = \theta - \theta_0$, K_π is the bulk modulus, $\beta_\pi = \beta_\pi(T, c_\pi)$ the thermal expansion coefficient and γ_π the chemical expansion coefficient.

The stress tensor in the solid phase is

$$t^s = (1 - n)(t_e^s - Ip^s) \tag{33}$$

where by neglecting the dependence of Helmholtz free energies on void fraction (Gray and Hassanisadeh 1991b) pressure in the solid phase is

$$p^s = p^w S_w + p^g S_g \tag{34}$$

and

$$\sigma' = (1 - n) t_e^s \tag{35}$$

is the effective stress tensor.

It can be easily shown from the above equations (Lewis and Schrefler 1998), that the total stress σ acting on a unit area of the multiphase medium is

$$\sigma = \sigma' - I(S_w p^w + S_g p^g) \tag{36}$$

The effective Jaumann rate of the stress is linked to the Eulerian strain rate tensor **D** by means of a constitutive relationship

$$\dot{\sigma}' = C_T(D^s - D_0^s) \tag{37}$$

where

$$C_T = C_T(D^s, \sigma', T) \tag{38}$$

is the tangent matrix, and D_0^s represents all other strains not directly associated with stress changes.

A particular form of C_T for partially saturated soils will be shown in section 6.3.

The solid density may be written as (Lewis and Schrefler 1998)

$$\frac{1}{\rho^s}\frac{\overset{s}{D}\rho^s}{Dt} = \frac{1}{1-n}\left[(\overline{\alpha}-n)\frac{1}{K_s}\frac{\overset{s}{D}p^s}{Dt} - \beta_s(\overline{\alpha}-n)\frac{\overset{s}{D}T}{Dt} - (1-\overline{\alpha})\mathrm{div}\ \overline{v}^s\right] \tag{39}$$

where

$$\overline{\alpha} = 1 - \frac{K_t}{K_s} \tag{40}$$

is Biot's constant, K_s the bulk modulus of the grain material and K_t the bulk modulus of the skeleton. For incompressible grain material $\overline{\alpha} = 1$. This is assumed here for the general model. This does not imply that the solid skeleton is rigid because of rearrangement of the voids. The necessary evolution equation for this is the mass balance equation for the solid (8).

Mass transport of fluid phases due to flow is governed by Darcy's law. This law is obtained from the linear momentum balance equation and is given in section 4, (Lewis and Schrefler 1998). A constitutive assumption is however necessary for the momentum exchange term of eq. (23). The dissipative part of the fluid-solid exchange of momentum is given by

$$\rho_\pi \hat{t}^\pi = -R^\pi \eta^\pi \overline{v}^{\pi\alpha} + p^\pi\ \mathrm{grad}\ \eta^\pi \tag{41}$$

By assuming that R^π is invertible

$$K^\pi = \eta^\pi (R^\pi)^{-1} = \frac{k\ k^{r\pi}}{\mu^\pi}(\rho^\pi,\eta^\pi,T) \tag{42}$$

where $\mu^\pi = \mu^\pi(\rho^\pi,T,c_\pi)$ is dynamic viscosity, k is intrinsic permeability tensor and $k^{r\pi} = k^{r\pi}(p^c,T)$ the relative permeability, accounting for dissipative terms at fluid-solid and fluid-fluid interfaces when several fluid phases are present. The intrinsic permeability depends solely on properties of the solid matrix (Bear and Bachmat 1984) and varies with the void ratio e

$$e = \frac{dv^w + dv^g}{dv^s} \tag{43}$$

Diffusive-dispersive mass flux is governed by Fick's law

$$J_\alpha^\pi = -\rho^\alpha D_\alpha^\pi\ \mathrm{grad}\left(\frac{\rho^{\pi\alpha}}{\rho^\alpha}\right) \tag{44}$$

where $\mathbf{D}_\alpha^\pi$ is the effective dispersion tensor, π is diffusing phase and α the phase in which diffusion takes place $(\alpha = w, g)$. $\mathbf{D}_\alpha^\pi$ is a function of the tortuosity factor, which accounts for the tortuous nature of the pathway in soil; because of mechanical dispersion, it is also correlated with seepage velocity (Manassero and Shakelford 1994).

For dry air and water vapour (binary system) we have in particular

$$J_g^{ga} = -\rho^g \mathbf{D}_g^{ga} \operatorname{grad}\left(\frac{\rho^{ga}}{\rho^g}\right)$$

$$J_g^{gw} = -\rho^g \mathbf{D}_g^{gw} \operatorname{grad}\left(\frac{\rho^{gw}}{\rho^g}\right)$$

(45)

Eq. (12) results in

$$\operatorname{grad}\left(\frac{\rho^{ga}}{\rho^g}\right) = \operatorname{grad}\left(\frac{\rho^g - \rho^{gw}}{\rho^g}\right) = -\operatorname{grad}\left(\frac{\rho^{gw}}{\rho^g}\right)$$

(46)

From equations (18) and (46) it follows, that

$$\mathbf{D}_g^{ga} = \mathbf{D}_g^{ga} = \mathbf{D}_g$$

(47)

Finally, using (29) we obtain for binary gas

$$J_g^{ga} = -\rho^g \frac{M_a M_w}{M_g^2} \mathbf{D}_g \operatorname{grad}\left(\frac{p^{ga}}{p^g}\right) = \rho^g \frac{M_a M_w}{M_g^2} \mathbf{D}_g \operatorname{grad}\left(\frac{p^{gw}}{p^g}\right) = -J_g^{gw}$$

(48)

The total heat flux $\mathbf{q}$ in the multiphase medium, sum of the partial heat fluxes $\mathbf{q}^\pi$, is governed by Fourier's law

$$\mathbf{q} = -\lambda_{\text{eff}} \operatorname{grad} T$$

(49)

where λ_{eff} is effective thermal conductivity tensor.

The saturation S_w is an invertible function of capillary pressure p^c and T

$$S_w = S_w\left(p^c, T\right)$$

(50)

For soils this relationship is obtained directly from laboratory tests. For building materials such a relationship is usually obtained via the Laplace equation

$$p^c = \frac{2\sigma}{r}$$

(51)

where $\sigma = \sigma(T)$ is surface tension and r pore size radius. The pore size distribution as a function of the saturation is obtained through experimental tests, such as centrifuge tests, sorption isotherm measurements or mercury porosimetry. Because of this way of determining r, eq. (51) can be considered as a macroscopic relationship. The case, where water is present as one or more molecular layers adsorbed on the surface of a solid because of the Van der Waals and/or other interactions will be dealt with below.

Finally, for the model closure we need thermodynamic relations (Baggio et al. 1997). For the relationship between relative humidity H and capillary pressure in the pores, Kelvin-Laplace law is assumed to be valid

$$H = \frac{p^{gw}}{p_s^{gw}} = \exp\left(-\frac{p^c M_w}{\rho^w R \theta} \right)$$

(52)

The water vapour saturation pressure p_s^{gw} is obtained from the Clausius-Clapeyron equation

$$p_s^{gw}(\theta) = p_{so}^{gw} \exp\left[-\frac{M_w \Delta h_{vap}}{R}\left(\frac{1}{\theta} - \frac{1}{\theta_0} \right) \right]$$

(53)

where θ_0 is a reference temperature, p_{so}^{gw} is water vapour saturation pressure at θ_0 and Δh_{vap} the specific enthalpy of evaporation. This equation is obtained from the second law of thermodynamics and is valid in the vicinity of θ_0.

Since capillary pressure p^c is equal to the water potential ψ multiplied by a constant $(p^c = -\rho_w \psi)$, eq. (52) can also be used below the capillary region (Baggio et al. 1997).

4   General Field Equations

The macroscopic balance laws are now transformed and constitutive equations introduced, to obtain the general field equations, which will then be solved numerically. Here only the results are given, for a full picture the reader is referred to (Lewis and Schrefler 1998).

The linear momentum balance equation for fluids after neglecting several terms gives the generalised Darcy equation

$$\eta^\pi \overline{v}^{\pi s} = \frac{k\, k^{r\pi}}{\mu^\pi}\left[-\operatorname{grad} p^\pi + \rho^\pi\left(g - \overline{a}^s - \overline{a}^{\pi s} \right) \right]$$

(54)

where $\overline{a}^{\pi s}$ is acceleration relative to the solid.

By summing the linear momentum balance equations for all constituents, by taking into account the definition of total stress eq. (36), and eq. (24), by assuming continuity of stress at the fluids-solid interfaces and by introducing the averaged density of the multiphase medium

$$\rho = (1-n)\rho^s + \sum_{\pi \neq s} n S_\pi \rho^\pi \tag{55}$$

we obtain its linear momentum balance equation.

For a multiphase medium, made of the solid, water and gas (mixture) constituents this equation is

$$\operatorname{div} \sigma + \rho(g - \overline{a}^s) - n S_w \rho^w \overline{a}^{ws} - n S_g \rho^g \overline{a}^{gs} = 0 \tag{56}$$

The sum of the macroscopic mass balance equation for solid (8) and for water (10) (to eliminate the time derivative of porosity n) gives the continuity equation for water

$$\frac{(1-n)}{\rho^s}\frac{\overset{s}{D}\rho_s}{Dt} + \operatorname{div}\overline{v}^s + \frac{n}{\rho_w}\frac{\overset{s}{D}\rho^w}{Dt} + \frac{n}{S_w}\frac{\overset{s}{D}S_w}{Dt} + \frac{1}{S_w\rho_w}\operatorname{div}\left(n S_w\rho^w\overline{v}^{ws}\right) = -\frac{1}{S_w\rho^w}\dot{m} \tag{57}$$

where Darcy's law eq. (5.4) has yet to be introduced.

Similarly the sum of the macroscopic mass balance equation for solid (8) and the mass balance equation for gas (54) gives the continuity equation for gas (mixture)

$$\frac{(1-n)}{\rho^s}\frac{\overset{s}{D}\rho^s}{Dt} + \operatorname{div}\overline{v}^s + \frac{n}{\rho^g}\frac{\overset{s}{D}\rho^g}{Dt} + \frac{n}{S_g}\frac{\overset{s}{D}S_g}{Dt} + \frac{1}{\rho^g S_g}\operatorname{div}\left(n S_g\rho^g\overline{v}^{gs}\right) = \frac{\dot{m}}{\rho^g S_g} \tag{58}$$

Without inertial terms and phase change these mass balance equations were used in (Schrefler et al. 1994) for pollutant transport analysis. In the form given here they allow distinction between pollutant transport in gaseous and liquid phases.

For heat transfer analysis in partially saturated porous media another form of the mass balance equations is more convenient. These will be derived next as well as the other balance equations under the following simplifying assumptions. We consider slow phenomena only, hence inertia terms disappear. Further we assume small strain situations, which allow consideration of partial instead of material time derivatives. The final form of equations is then obtained for a multiphase system of solid, gas, water (or immiscible pollutant) as follows (Lewis and Schrefler 1998).

The vapour mass balance eq. (15) is subtracted from gas mass balance eq. (11) to obtain dry air mass balance equation. This allows elimination of the mass rate of water evaporation. Furthermore Darcy's law eq. (54) is introduced, where the body force is neglected, and Fick's law in the form of eq. (48)

$$\frac{\partial}{\partial t}\left[n S_g\rho^{ga}\right] + S_g\rho^{ga}\operatorname{div}\overline{v}^s$$

$$- \operatorname{div}\left(\rho^{ga}\frac{k\,k^{rg}}{\mu^g}\operatorname{grad}p^g\right) + \operatorname{div}\left(\rho^g\frac{M_a M_w}{M^2}D_g\operatorname{grad}\left(\frac{p^{gw}}{p^g}\right)\right) = 0 \tag{59}$$

The vapour mass balance eq. (15) and the water mass balance eq. (10) are now summed, to obtain a mass balance equation for all water species.

Again $\dot{m}$ disappears. Further Darcy's law and Fick's law are introduced as above, yielding

$$\frac{\partial}{\partial t}\left[n\,S_g\rho^{gw}\right]+S_g\rho^{gw}\mathrm{div}\,\overline{v}^s$$

$$-\mathrm{div}\left(\rho^{gw}\frac{k\,k^{rg}}{\mu_g}\mathrm{grad}\,p^g\right)-\mathrm{div}\left(\rho^g\frac{M_aM_w}{M^2}D_g\,\mathrm{grad}\left(\frac{p^{gw}}{p^g}\right)\right)= \tag{60}$$

$$=-\frac{\partial}{\partial t}\left(nS_w\rho^w\right)-S_w\rho^w\mathrm{div}\,\overline{v}^s+\mathrm{div}\left(\rho^w\frac{k\,k^{rw}}{\mu^w}\left(\mathrm{grad}\,p^g-\mathrm{grad}\,p^c-\rho^w\mathbf{g}\right)\right)$$

The energy balance equation for the whole multiphase medium is obtained by summation of eq. (25) for all constituents. Fourier's law eq. (49) is applied as well as enthalpy definition. This allows the energy balance to be rewritten in terms of heat capacity (Lewis and Schrefler 1998). Heat capacity of the multiphase medium at constant pressure

$$\rho C_p = n\,S_w\rho^w C_p^w + n\,S_g\rho^g C_p^g + (1-n)\rho^s C^s \tag{61}$$

results from averaging of the single phase heat capacities $C_p^\pi = C_p^\pi(T, c^\pi)$. Finally the mass rate of evaporation is eliminated through the water phase mass balance equation and convective velocities are substituted by Darcy's law (54), yielding

$$\rho C_p\frac{\partial T}{\partial t}-\mathrm{div}\left(\lambda_{eff}\,\mathrm{grad}\,T\right)$$

$$-\left[C_p^w\rho^w\frac{kk^{rw}}{\mu^w}\left(\mathrm{grad}\,p^g-\mathrm{grad}\,p^c-\rho^w\mathbf{g}\right)+C_p^g\rho^{gw}\frac{kk^{rg}}{\mu^g}\mathrm{grad}\,p^g\right]\cdot\mathrm{grad}\,T= \tag{62}$$

$$= \Delta h_{vap}\left[\frac{\partial}{\partial t}\left(nS_w\rho^w\right)+S_w\rho^w\mathrm{div}\,\overline{v}^s-\mathrm{div}\left(\rho^w\frac{kk^{rw}}{\mu^w}\left(\mathrm{grad}\,p^g-\mathrm{grad}\,p^c-\rho^w\mathbf{g}\right)\right)\right]$$

The linear momentum balance equation (56) for the particular multiphase medium under consideration after introduction of eq. (36) and the constitutive relationship (37) and substituting $\mathbf{D}$ by ε dt because of small strain assumption becomes,

$$\mathrm{div}\left[\left(C_T\left(d\varepsilon-I\frac{\beta_s}{3}dT-d\varepsilon^\circ\right)\right)-I\left(p_g-S^wp^c\right)\right]+\rho\mathbf{g}=0 \tag{63}$$

where ε is the linear strain tensor, $I\frac{\beta_s}{3}dT$ strain caused by thermoelastic expansion and ε_0 represents all other strains not directly associated with stress changes.

In the presence of an immiscible pollutant we must consider its mass balance equation as well as the above equations

$$\frac{\partial\left(\eta^{\pi}\rho^{\pi}\right)}{\partial t}+\frac{\partial\left[(1-n)\rho^{s}\right]}{\partial t}+\operatorname{div}\left(\eta^{\pi}\rho^{\pi}\overline{v}^{s}\right)$$
$$+\operatorname{div}\left[\rho^{\pi}\frac{k\,k^{r\pi}}{\mu^{\pi}}\left(\operatorname{grad}p^{\pi}-\rho^{\pi}\,\mathbf{g}\right)\right]+\operatorname{div}\left[(1-n)\rho^{\pi}\overline{v}^{s}\right]=0 \tag{64}$$

In the case of contaminants soluble in water the solute conservation equation, after introduction of Fick's law (44) and Darcy's law (54) is

$$\frac{\partial}{\partial t}\left(n\,S_{w}c_{w}^{\pi}\right)-\operatorname{div}\left[c_{w}^{\pi}\frac{k\,k^{rw}}{\mu^{w}}\left(\operatorname{grad}p^{w}-\rho^{w}g\right)\right]=\operatorname{div}\left(\rho^{\pi}\mathbf{D}_{w}^{\pi}\operatorname{grad}c_{w}^{\pi}\right)+I_{w} \tag{65}$$

Similarly, the solute conservation equation in gas may be written as

$$\frac{\partial}{\partial t}\left(n\,S_{g}c_{g}^{\pi}\right)-\operatorname{div}\left[c_{g}^{\pi}\frac{k\,k^{rg}}{\mu^{g}}\left(\operatorname{grad}p^{g}\right)\right]=$$
$$\operatorname{div}\left(\rho^{\pi}\mathbf{D}_{g}^{\pi}\operatorname{grad}c_{g}^{\pi}\right)+I_{g} \tag{66}$$

All saturations of the fluids present must sum to one.

4.1 Initial and Boundary Conditions

Furthermore it is necessary to define the initial and boundary conditions. The initial conditions specify the full fields of gas pressure, capillary or water pressure, temperature, displacements, pressure of immiscible pollutant and concentrations according to constituents present and primary variables chosen; this last aspect will be specified, when the different models are presented.

$$p^{g}=p_{0}^{g},p_{c}=p_{0}^{c},\ T=T_{0},\ \mathbf{u}=\mathbf{u}_{0},\ p^{\pi}=p_{0}^{\pi},\ c^{\pi}=c_{0}^{\pi},\ \text{at}\ t=0. \tag{67}$$

The boundary conditions can be imposed values on Γ_{π} or fluxes on Γ_{π}^{q}, where the boundary $\Gamma=\Gamma_{\pi}\cup\Gamma_{\pi}^{q}$. The imposed values on the boundary for gas pressure, capillary or water pressure, temperature, displacements, immiscible pollutant pressures and concentrations are as follows

$$p^{g}=\hat{p}^{g}\ \text{on}\ \Gamma_{g},\quad p^{c}=\hat{p}^{c}\ \text{on}\ \Gamma_{c},\ T=\hat{T}\ \text{on}\ \Gamma_{T},\ \mathbf{u}=\hat{\mathbf{u}}\ \text{on}\ \Gamma_{u}$$

$$p^{\pi}=\hat{p}^{\pi}\text{on}\ \Gamma_{\pi}\ \text{(for immiscible pollutant)} \tag{68}$$

$c_w^\pi = \hat{c}_w^\pi$ on $\Gamma_{\pi w}$, $c_g^\pi = \hat{c}_g^\pi$ on $\Gamma_{\pi g}$ (for miscible pollutant)

The volume averaged flux boundary conditions for water species and dry air conservation equations and the energy equation, to be imposed at the interface between the porous media and the surrounding fluid are as follows

$$
\left(\rho^{ga}\overline{\mathbf{v}}^g - \rho^g\,\overline{\mathbf{v}}^{gw}\right)\mathbf{n} = q^{ga} \quad \text{on } \Gamma_g^q,
$$

$$
\left(\rho^{gw}\overline{\mathbf{v}}^g + \rho^w\overline{\mathbf{v}}^w + \rho^g\overline{\mathbf{v}}^{gw}\right)\mathbf{n} = \beta_c\left(\rho^{gw} - \rho_\infty^{gw}\right) + q^{gw} + q^w \text{ on } \Gamma_c^q, \tag{69}
$$

$$
-\left(\rho^w\overline{\mathbf{v}}^w\Delta h_{vap} - \lambda_{eff}\nabla T\right)\mathbf{n} = \alpha_c\left(T - T_\infty\right) + q^T \text{ on } \Gamma_T^q,
$$

where $\mathbf{n}$ is the unit vector, perpendicular to the surface of the porous medium, pointing toward the surrounding gas, ρ_∞^{gw} and T_∞ are, respectively, the mass concentration of water vapour and temperature in the undisturbed gas phase distant from the interface, α_c and β_c are convective heat and mass transfer coefficients, while q^{ga}, q^{gw}, q^w and q^T are the imposed dry air flux, imposed vapour flux, imposed liquid flux and imposed heat flux respectively.

Further we have for immiscible pollutant

$$
\left(\rho^\pi\overline{\mathbf{v}}^\pi\right)\mathbf{n} = \mathbf{q}^\pi \quad \text{on } \Gamma_\pi^q \tag{70}
$$

and for miscible pollutant

$$
\begin{aligned}
\left(c_w^\pi\overline{\mathbf{v}}^{ws} - \mathbf{D}_w^\pi \operatorname{grad} c_w^\pi\right)\mathbf{n} = q^{\pi w} \quad &\text{on } \Gamma_{\pi w}^q \\
\left(c_g^\pi\overline{\mathbf{v}}^{gs} - \mathbf{D}_g^\pi \operatorname{grad} c_g^\pi\right)\mathbf{n} + q^{\pi g} \quad &\text{on } \Gamma_{\pi g}^q
\end{aligned} \tag{71}
$$

where $q^\pi, q^{\pi w}$ and $q^{\pi g}$ are the imposed fluxes.

Equations (69-71) are the natural boundary conditions, respectively, for the dry air conservation equation (59), water species conservation equation (60) and energy conservation equation (62), when the solution of these equations is obtained through a weak formulation of the problem, as usually done with the finite element method.

The traction boundary conditions for the displacement field are

$$
\sigma \cdot \mathbf{n} = \mathbf{t} \quad \text{on } \Gamma_u^q, \tag{72}
$$

where $\mathbf{t}$ is the imposed traction.

5   The Numerical Model for Contaminant and Heat Transport in Deforming Porous Media

In the previous section a rather general mathematical model has been developed for the case of coupled thermo-hydro-mechanical problems in porous bodies, including contaminant transport. In this section the numerical solution of the full set of equations will be shown. In particular we deal with slow transient phenomena involving heat transfer and contaminant transport in deforming partially saturated porous media. For dynamics problems which are important in the simulation of natural hazards, the reader is referred to (Lewis and Schrefler 1998, Zienkiewicz et al. 1999, Meroi and Schrefler 1995).

According to the fact that contaminants are soluble in fluid phases or immiscible, the model will consist of a different number and type of equations. In the presence of a non-aqueous phase liquid NAPL, equations (57), (58) after introduction of (54) and without inertia terms and equations (62), (63) and (64) are solved, while in the presence of soluble contaminants eq. (65) and (66) substitute equation (64). Together with the governing equations the pertinent constitutive relationships must be considered. In this case the following primary variables have been chosen (Schrefler et al. 1994, Schrefler et al. 1997): displacements of the solid skeleton, water and gas pressures and temperatures; the variables associated with the contaminant depend on the particular problem. The choice is however between concentration and phase pressure.

5.1 Discretisation in Space

A weak formulation of the governing equations is obtained by applying Galerkin's procedure of weighted residuals (Zienkiewicz and Taylor 1989). Terms involving second spatial derivatives are transformed by means of Gauss's theorem. Then field variables are approximated in space as is usual in finite element techniques and expressed in terms of their nodal variables. The equations to be accounted for depend on the pollutant considered, for instance in the case of soluble contaminant, assuming that no interphase change or source terms are present, the semi-discretised system of ordinary differential equations in time takes the following matricial form

$$\int_{\Omega} \mathbf{B}^T \dot{\sigma}' d\Omega + \mathbf{C}_{sw}\dot{\mathbf{p}}^w + \mathbf{C}_{sa}\dot{\mathbf{p}}^g + \mathbf{C}_{st}\dot{\mathbf{T}} = \dot{\mathbf{f}}_s$$

$$\mathbf{C}_{ws}\dot{\mathbf{u}} + \mathbf{P}_{ww}\dot{\mathbf{p}}^w + \mathbf{C}_{wa}\dot{\mathbf{p}}^g + \mathbf{C}_{wt}\dot{\mathbf{T}} + \mathbf{C}_{wc_w}\dot{\mathbf{C}}_w^\pi + \mathbf{C}_{wc_a}\dot{\mathbf{C}}_g^\pi + \mathbf{H}_{ww}\mathbf{p}^w = \dot{\mathbf{f}}_w$$

$$\mathbf{C}_{as}\dot{\mathbf{u}} + \mathbf{C}_{aw}\dot{\mathbf{p}}^w + \mathbf{P}_{aa}\dot{\mathbf{p}}^g + \mathbf{C}_{at}\dot{\mathbf{T}} + \mathbf{C}_{ac_w}\dot{\mathbf{C}}_w^\pi + \mathbf{C}_{ac_a}\dot{\mathbf{C}}_g^\pi + \mathbf{H}_{aa}\mathbf{p}^g = \dot{\mathbf{f}}_a \qquad (73)$$

$$\mathbf{C}_{tw}\dot{\mathbf{p}}^w + \mathbf{C}_{ta}\dot{\mathbf{p}}^g + \mathbf{P}_{tt}\dot{\mathbf{T}} + \mathbf{C}_{tc_w}\dot{\mathbf{C}}_w^\pi + \mathbf{C}_{tc_a}\dot{\mathbf{C}}_g^\pi + \mathbf{H}_{tt}\mathbf{T} = \dot{\mathbf{f}}_t$$

$$\mathbf{C}_{c_w w}\dot{\mathbf{p}}^w + \mathbf{C}_{c_w a}\dot{\mathbf{p}}^g + \mathbf{C}_{c_w t}\dot{\mathbf{T}} + \mathbf{P}_{c_w c_w}\dot{\mathbf{C}}_w^\pi + \mathbf{C}_{c_w c_a}\dot{\mathbf{C}}_g^\pi + \mathbf{H}_{c_w c_w}\mathbf{C}_w^\pi = \dot{\mathbf{f}}_{c_w}$$

$$\mathbf{C}_{c_a w}\dot{\mathbf{p}}^w + \mathbf{C}_{c_a a}\dot{\mathbf{p}}^g + \mathbf{C}_{c_a t}\dot{\mathbf{T}} + \mathbf{C}_{c_a c_w}\dot{\mathbf{C}}_w^\pi + \mathbf{P}_{c_a c_a}\dot{\mathbf{C}}_g^\pi + \mathbf{H}_{c_a c_a}\mathbf{C}_g^\pi = \dot{\mathbf{f}}_{c_a}$$

The matrices are listed in the Appendix. Equations (73) represent a coupled non-symmetric and non-linear system, which needs to be integrated in time. This system can be written in concise form as

$$\mathbf{C}(\mathbf{x})\dot{\mathbf{x}} + \mathbf{K}(\mathbf{x})\mathbf{x} = \dot{\mathbf{F}}(\mathbf{x}) \qquad (74)$$

where $\mathbf{x}^T = (\mathbf{u}, \mathbf{p}^w, \mathbf{p}^g, \mathbf{T}, \mathbf{c}_w^\pi, \mathbf{c}_g^\pi)$ and matrices $\mathbf{C}$, $\mathbf{K}$ and $\dot{\mathbf{F}}$ are obtained from eq. (73) by inspection.

Note that here the equilibrium equation has been differentiated with respect to time. The ensuing integration in the time domain of the system allows us to also introduce non linear behaviour of the solid skeleton (Lewis and Schrefler 1998). In this procedure the first term of the first equation in (73) is substituted by the tangent stiffness matrix $\mathbf{K}_T$. For dealing with elastoplastic solid behaviour, the reader is referred to (Lewis and Schrefler 1998).

In the following only the direct or monolithic solution of the system of equations ensuing from integration in time domain will be considered. An alternative to this procedure is the staggered procedure, where after operator splitting, iterations between subsets of equations are carried out. The interested reader is referred e.g. to (Lewis and Schrefler 1998, Park and Felippa 1983, Turska and Schrefler 1993).

5.2 Discretisation in the Time Domain

We assume consistency and convergence of the finite element discretisation in space (Zienkiewicz and Taylor 1989). Using a one-step finite difference operator for the time derivative, we obtain the discrete equation for (73)

$$\mathbf{A}\,\mathbf{x}_{n+1} - \mathbf{B}\,\mathbf{x}_n - \mathbf{F} = 0 \tag{75}$$

where, for instance, for the generalised mid-point method

$$\begin{aligned}
\mathbf{A} &= \mathbf{C} + \vartheta \Delta t \mathbf{K} \\
\mathbf{B} &= \mathbf{C} - (1 - \vartheta) \Delta t \mathbf{K} \\
\mathbf{F} &= \dot{\mathbf{F}} \Delta t
\end{aligned} \tag{76}$$

and ϑ is a parameter usually in the range [0,1]. Since the problem is non-linear, $\mathbf{A}$ and $\mathbf{B}$ depend on $\mathbf{x}$.

We evaluate here $\mathbf{x}_{n+1}$ directly from eq. (75) by means of a fixed-point type procedure

$$\mathbf{x}_{n+1,k+1} = \mathrm{f}\!\left(\mathbf{x}_{n+1,k}\right) \tag{77}$$

Although this procedure may also be used for elastoplastic skeleton behaviour as mentioned above, in that instance a Newton Raphson type procedure is advisable. For this and for the error analysis the interested reader is referred to (Lewis and Schrefler 1998).

6 Numerical Examples

Three examples follow to show the wide range of applicability of the developed model. The first one which deals with transport of a dense non aqueous phase liquid in a deforming soil, was solved by Schrefler et al. (1994) and by Lewis et al. (1998). The second considers pollutant transport in water and gas phases (Zhan et al. 1995). The third one deals with reservoir compaction in exploited gas reservoirs, where the effects of capillary pressure, eq. (31), are investigated. An appropriate constitutive relationship for the solid skeleton accounting for capillary effects is shown in this instance.

6.1 Immiscible Pollutant

Dense non-aqueous phase liquids (DNAPL), such as chlorinated solvents, polychlorinated biphenyl oils and creosotes, exist as separated fluid phase in ground water and are classified into a unique class of subsurface hazardous water pollutants. Because these compounds are relatively inexpensive to produce and have a wide variety of uses, the great potential of pollution in ground water has become an environmental concern (Parker 1989). In recent years, a large number of numerical simulators have been developed to model NAPL transport in multi-phase systems, e.g. (Abriola and Pinder 1985, Corapcioglu and Baehr 1987, Kueper and Frind 1989, Mendoza and Frind 1990).

However coupling of transport with soil deformation has received very little attention up to date. In Schrefler et al. (1994), based on the general coupled model of sections 5.1, a numerical simulation of the water drainage from and the infiltration of DNAPL into a porous medium was presented. The problem was then solved also by Lewis et al. (1998). Some results are shown here.

The test example involves a vertical column of 1 m in height with a soil permeability of $1 \cdot 0 \times 10^{-11}$ m^2 and $1 \cdot 5 \times 10^{-12}$ m^2. The source boundary condition was set at $0 \cdot 5$ of the NAPL fluid saturation and a porosity of $0 \cdot 40$ was used throughout the column, initially completely saturated by water. The result is a displacement of the water from the column by the encroaching NAPL. The parameters chosen for the simulations were a NAPL density of 1461 kg/m^3 and a viscosity of $0 \cdot 57 \times 10^{-3}$ N s/m^2. The water phase was assigned a density of 1000 kg/m^3 and a viscosity of $1 \cdot 0 \times 10^{-3}$ N s/m^2. The resulting profiles of vertical displacements and effective water saturation are shown in Figure 1(a) and (b), respectively, for different time values. The results show the influence of permeability of the medium for saturation and vertical displacement.

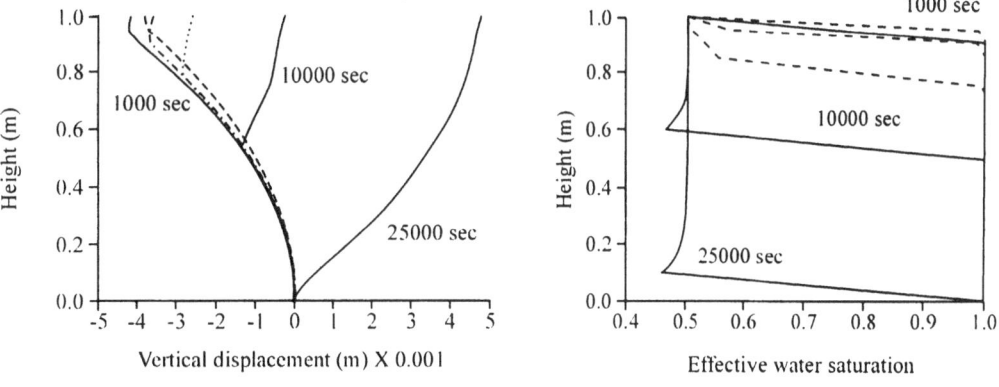

Figure 1. Vertical displacements and effective water saturation for different time values (Lewis et al. 1998).

6.2 Multiphase Flow, Heat Flow and Solute Transport in Deforming Porous Media

As a second example, we analyse pollutant transport in water and air phases (Zhan et al. 1995). A similar problem has been solved by Dakshanamurthy and Fredlund (1981). These authors studied the moisture and heat flow in a column of unsaturated subgrade soil, for instance below a highway or airfield pavement, due to the effect of environmental changes. The subgrade system is assumed initially in a equilibrium state. Then sudden environmental changes, e.g. evaporation, infiltration and/or temperature variation, are imposed at the boundary, either positive or negative with respect to the initial equilibrium state. Similarly, temperature variation results in a gradient at the surface and an overpressure. Consequently simultaneous water and air phases flow takes place. Here also the associated convective/diffusive pollutant transport is considered (Zhan et al. 1995).

Some differences exist between the two models, both from the theoretical and numerical point of view. Dakshanamurthy and Fredlund firstly solved by an explicit method the heat flow equation, then solved by an explicit forward difference technique the transient flow equation of water and air respectively. Moreover they used a simpler set of equations: no water flow depending on temperature gradient was taken into account, resulting in fact that water content distribution in space and time is essentially the same for isothermal and non-isothermal conditions. In addition, constant permeability for water and air phases were assumed, soil deformation and solute transport were not considered. Here the problem is solved as coupled taking into account these effects. The relationships between relative permeability, saturation and capillary pressure proposed by Brooks and Corey (1966) are used. Solute transport in both fluid phases is considered. Except for the contaminant solid deformation, all simulation cases (say different boundary conditions and physical parameters) follow the paper by Dakshanamurthy and Fredlund (1981).

The problem is one-dimensional, but is modelled as a two-dimensional one by 10 square finite elements in vertical direction 0.1 m sided. Nine-node elements are used, each node having seven degrees of freedom. Temporal discretisation is fully implicit.

The parameters for soil are as follows: Young modulus $1*10^7$ Pa for deformable case and $1*10^{17}$ Pa for the undeformable one, Poisson ratio is 0.2, porosity 0.5, bulk modulus $6*10^{13}$ Pa and density 1800 Kg/m^3. Specific heat for soil is $1.25*10^5$ J/Kg°C and thermal expansion coefficient $1*10^4$ °C^{-1}. Absolute permeability is $0.22*10^{-14}$ m^2 for infiltration case and $0.57*10^{-14}$ m^2 for evaporation case. Parameters for water are the following: bulk modulus $8*10^{14}$ Pa, viscosity $1*10^{-3}$ Ns/m^2, specific heat 0 J/Kg°C and density 1000 Kg/m^3. Thermal expansion coefficient of water is 0 °C^{-1} and dispersivity $1*10^{-8}$ m^2/s. For air, the state equation of perfect gas is used, viscosity is assumed $1*10^{-5}$ Ns/m^2, specific heat 0 J/Kg°C, density 1.22 Kg/m^3 and air dispersivity $1*10^{-8}$ m^2/s. The heat conductivity of the medium is assumed equal to 1.15 W/m°K. Initial conditions for solutes in water and air phase are all zero everywhere and boundary conditions are unit constant water and air concentration at the top of the column for both infiltration (swelling) and evaporation (consolidation) cases. All other initial and boundary conditions are as in (Schrefler et al. 1995). Computational time step is initially one hour and then increased by one order after ten steps until 1000 hours. Simulation results demonstrate that the model shows good promise for describing the behaviour of unsaturated soil systems under highly complex environmental changes.

Figure 2 shows the comparison of the temperature isotherms within the soil layer as a results of an increase in the temperature from 10 °C (283.2 °K) to 25 °C (298.2 °K) under infiltration condition. The Figure illustrates how the imposed thermal gradient at the surface slowly proceeds toward the bottom of the sample and eventually a new equilibrium state corresponding to the new boundary conditions is obtained. Comparison is shown in Figure 2 between results from Dakshanamurthy and Fredlund (dotted lines with indicated times) and present solution.

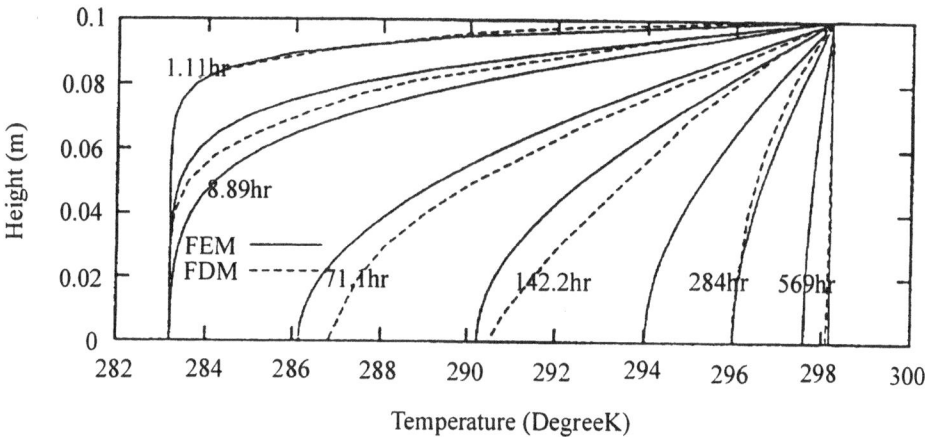

Figure 2. Profiles of temperature by Dakshanamurthy and Fredlund (dotted lines) and by present model during swelling.

In the case of volume increase (swelling) due to infiltration, Figure 3 shows the corresponding air pressure distributions for both deformable and undeformable assumptions. It is noted that air pressure builds up stronger for the undeformable than for deformable situation.

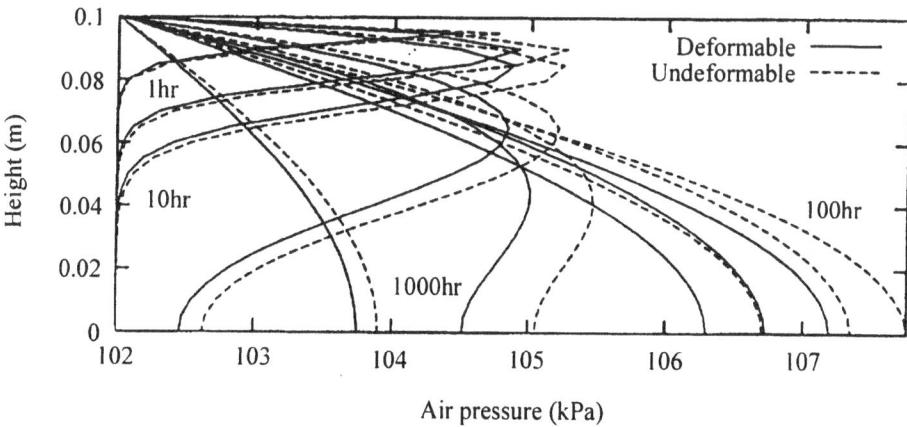

Figure 3. Profiles of air pressure during swelling for deformable (solid lines) and undeformable (dotted lines) soil.

The swelling process due to infiltration is shows in Figure 4, when assuming deformable soil.

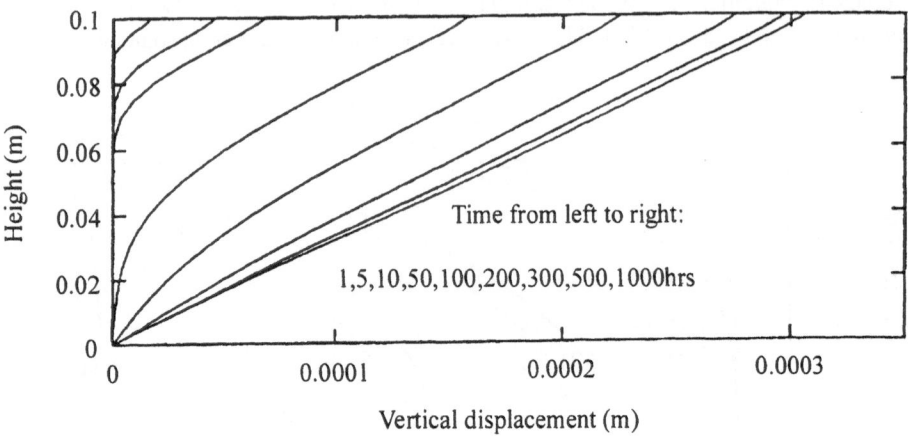

Figure 4. Profiles of vertical displacements during swelling.

Pore water pressure distribution through the soil layer due to a change in pore water pressure at the boundary and the distribution of moisture content can be found in Schrefler et al. (1995). In the same paper, results for the reverse processes (evaporation), were also presented in detail. For the pollutant transport, due to the boundary conditions imposed at the top of the column, it is expected a downward movement due to dispersion mechanism and advection. Figures 5a-b show this movement in water and air for both consolidation (dotted lines) and swelling (solid lines).

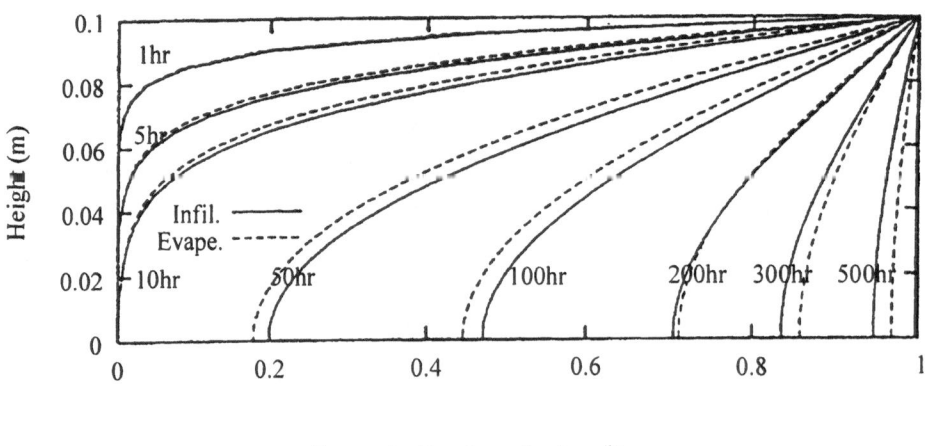

Figure 5a. Profile of concentrations in water for both swelling (solid line) and consolidation cases (dotted line).

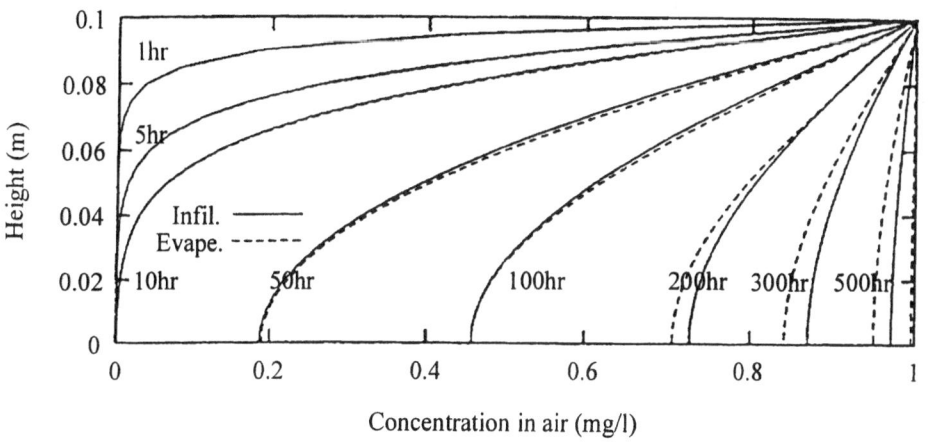

Figure 5b. Profile of concentrations in air for both swelling (solid line) and consolidation cases (dotted line).

It is noted that for swelling case, corresponding to infiltration, concentration in water in early time is higher than in the consolidation case. This is due to evaporation because of stronger advective inflow. The opposite can be seen for pollutant movement in air. This fact is more pronounced in later times than in early times.

6.3 Reservoir Compaction in Exploited Gas Reservoirs

In reservoir engineering applications, compaction of gas bearing strata is usually analysed assuming from field measurements the gas pressure history and using a limited subset of the governing equations (73) previously presented (Lewis and Schrefler 1998). Real gas reservoirs, however, are made up of media (more or less cemented sands, porous rocks etc.) with an open structure, i.e. with many interconnected voids filled with gas and water. Observed degrees of water saturation in gas fields vary between 20-65% before exploitation and 65-85% at the depletion of the wells. Gas must have a continuous distribution in the zone of the wells otherwise it remains trapped in the reservoir. At the fluid interface, tensions occur which cause capillary effects or suction and strongly affect the mechanical behaviour of the solid skeleton. For sands in particular, the original geometric structure is maintained also by capillary effects, which result in apparent cohesion. When these decrease, the system assumes a different geometric configuration with a re-arrangement of grains or local fracture of rocks, both causing a reduction in void volume. The original system is said to be metastable. Capillary effects diminish when water saturation increases due to water flooding of the reservoir. During this transient phase, the areal extent of the discontinuity surfaces decreases and mechanical strength is reduced. In this situation a soil sample is said to collapse, with a macroscopic strain related to volume reduction. In the presence of capillary effects a stress measure has to be introduced which differs from the case where pores are fully saturated by the same fluid. Whereas in fully saturated media mechanical behaviour is controlled by one stress dimension variable (usually the effective stress), for a partially saturated medium two independent stress dimension parameters (combinations of total stress tensor, water and gas pressure) are needed to describe deformational behaviour. In the following, the assumed stress variables are suction, i.e. the difference between gas and water pressure, eq. (31), and Bishop's effective stress (36). This stress measure takes the form

$$\sigma'_{ij} = \sigma_{ij} - \left[S_r p_w + (1 - S_r) p_g\right] \delta_{ij} = \sigma_{ij} - \left[p_g - S_r s\right] \delta_{ij} \tag{78}$$

Soil suction s and water saturation may be mutually related by means of the following relationship (Alonso et al. 1990)

$$S_r = 1 - m \tanh(ls) \tag{79}$$

where m and l are material parameters. From equation (79) it can be seen that $S_r=1-m$ represents irreducible saturation, i.e. the limiting value of S_r as suction approaches infinity. In equation (79), parameter S_r represents a phenomenological measure of the capillary effects, through its experimental relationship with suction. Bishop's stress definition recovers the effective stress definition when saturation equals one, hence the consistency condition between stress measures is guaranteed.

An important consequence of the assumed definition is that changes in effective stress may occur, causing deformation to take place, which is related only to changes in water saturation (or capillary pressure). This mechanism is linked to water filtration through the solid skeleton. Hence it depends on water permeability and develops during long transient phases. Water inflow continues for some years after shutdown of the wells, hence saturation and Bishop's stress change continuously during these time spans. This may be one of the causes of ongoing subsidence which has in fact been observed.

Extensive laboratory experiments for collapsible geomaterials under oedometric and triaxial isotropic compression with variable saturation (suction) show a strong plastic behaviour involving volume strains caused by suction changes (Figure 6). With increasing saturation, the lowering of the yield limit is in the order of some megapascals (Delage et al. 1996), depending on the radius of capillary menisci, i.e. on the structure and dimension of pores.

It is also known that in sands and some types of rocks deformability increases with increasing water saturation, reaching a maximum in fully saturated conditions. This behaviour has been observed in clays, silts, clayey sands and also in chalk and sandstone. An initial nearly-elastic phase is shown, then a hardening plastic interval and an approximately linear behaviour associated with a decrease in the mean effective pressure (unloading). The nearer to its yield limit the material is, the stronger the influence of changes in saturation.

On the grounds of the mechanical behaviour described, to evaluate reservoir compaction and surface subsidence completely we must take into account the volumetric strain associated with changes in water saturation and the appropriate constitutive relationships have to be stated in an elastoplasticity frame. To this end we make use of the material model (Bolzon et al. 1996) where plastic volumetric behaviour dependent on suction changes is accounted for. The main features are now summarised as follows.

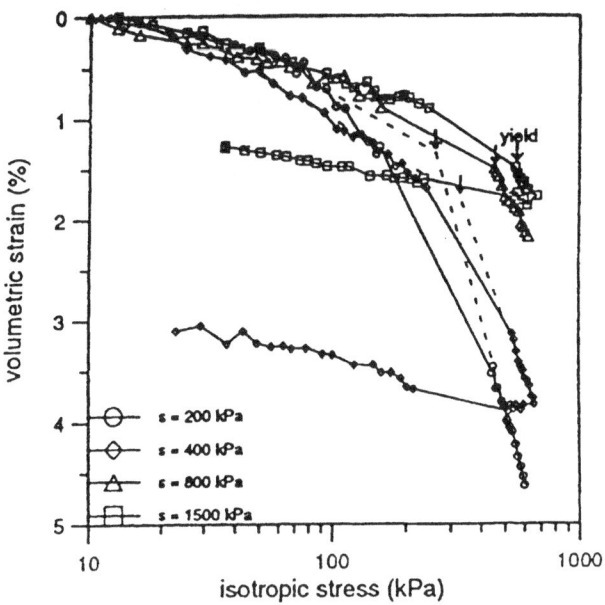

Figure 6. Volumetric strain of a silt sample under isotropic compression at different levels of suction (Delage et al. 1996).

To handle the previously described mechanical behaviour we refer to the generalised plasticity model (Pastor et al. 1990), in which:

the strain parameter involved in plasticity is the specific volume change. Recalling that specific volume v is the ratio between the sample volume and that of the solid, i.e. $v=1+e$, the strain parameter is defined as

$$\varepsilon_v = \frac{v - v_0}{v_0} = \frac{e - e_0}{1 + e_0} = \frac{n - n_0}{1 - n} \tag{80}$$

where e is the void ratio, n the porosity and subscript 0 refers to initial values;

total volumetric strain (recoverable and not) associated with the change in the hydrostatic component of the effective stress tensor p' can be written as

$$d\varepsilon_v = \frac{\lambda}{v_0} \frac{dp'}{p'} \tag{81}$$

soil compressibility being λ;

for an isotropic compression path, the volumetric plastic flow can be cast in the form

$$d\varepsilon_v^p = \frac{1}{H}\frac{dp'}{p'} \tag{82}$$

being plasticity not associated and H the hardening modulus;

hardening modulus, during an isotropic compression, is related to the slopes of the virgin loading path λ and elastic unloading line κ in the plane $(e, \ln p')$ by the equation

$$H = H_0 = \frac{1+e_0}{\lambda - \kappa} \tag{83}$$

along a generic stress path (isotropic or not), hardening modulus H_0 is substituted by

$$H = H_v(\eta)H_0 \tag{84}$$

$H_v(\eta)$ being a function depending on stress path, equal to one in the isotropic case.

The following changes are introduced in the model to account for dependence on suction along an isotropic compression (Bolzon et al. 1996):

a multiplicative representation of the hardening modulus is introduced, which in its simplest form linearly relates changes in stiffness and changes in suction, as

$$H = H_w(s)H_0 = (1+as)H_0 \tag{85}$$

where a is a material parameter. The volumetric plastic flow is still represented by equation (82), where p' now represents the mean Bishop's stress;

the dependence of soil compressibility on suction can be written in a form similar to equation (83) as

$$\lambda(s) = \frac{1+e_0}{(1+as)H_0} + \kappa \tag{86}$$

the change in total volumetric strain (sum of the reversible and irreversible parts) is given by

$$d\varepsilon_v = \frac{\lambda(s)}{v_0}\frac{dp'}{p'} \tag{87}$$

disregarding the elastic contribution, the compressibility modulus during plastic loading can be approximated as

$$\lambda(s) = \frac{\lambda(0)}{1+as} \tag{88}$$

$\lambda(0)$ being the compressibility of the saturated soil. This very simple approximated relation has the advantage of being very efficient computationally. A two-parameter exponential expression alternative to (88) can be found in the literature (Alonso et al. 1990);

the initial yield limit of the fully saturated case is modified in partially saturated conditions to introduce its dependence on suction as

$$p'_{y_i}(s) = p'_{y0_i} + is \tag{89}$$

where p'_{y0_i} represents the initial yield limit in saturated conditions. i has to be determined, e.g. by interpolation of experimental data to obtain an increasing function of suction when water saturation is less than one. From thermodynamic conditions, i must be greater than or equal to 1;

irreversible volumetric strain is assumed as the parameter controlling hardening, hence evolution of the yield surface. From equations (87)-(89), evolution of the yield surfaces is given as

$$p'_y(s) = \left(p'_{y0_i} + is\right)\left(\frac{p'_{y0}}{p'_{y0_i}}\right)^{1+as} \tag{90}$$

We now calculate, as an example the effect of plastic volumetric strain related to changes in saturation and pressure for the gas reservoir extensively analysed in (Lewis and Schrefler 1998, Simoni et al.) and shown in Figure 7. For simplicity we refer to the area in the neighbourhood of the symmetry axis where maximum compaction occurs and oedometric conditions prevail. As a consequence, we simply use 1-D conditions.

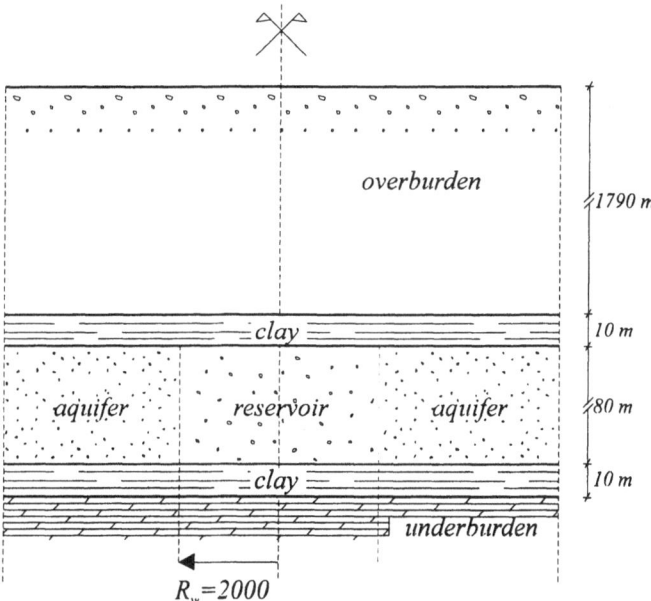

Figure 7. Cross-section and position of the ideal axisymmetric reservoir.

For the reservoir we assume an average vertical stress σ'_V of 22.8 MPa, horizontal stress σ'_H being $0.33\sigma'_V$, in accordance with oedometric conditions and used Poisson's ratio. The initial value of the mean stress p' is hence 12.7 MPa. At the intermediate depth of 1840 m below the surface, water pressure is about 19.5 MPa, whereas gas pressure is 19.7 MPa. Soil suction and water saturation are related by equation (79), where $m=0.9$ and $l=2$ MPa^{-1} are assumed. As a consequence of the hypotheses used, the initial water saturation is 0.65. The initial porosity, consistent with gas contained in the reservoir, is 17.1%, corresponding to the specific volume $v_0=1.2$, soil compressibility in fully saturated conditions $\lambda(0)$ is 0.045, the slope of initial yield surface is 1 and material parameter a of equation (88) is assumed as 5 MPa. The initial yield limit in saturated conditions p'_{y0_i} (equation (90)) is therefore 12.5 MPa. All these data are in accordance with field observations (AGIP 1996).

First of all, we analyse the effect of a reservoir pressure drop of 2 MPa. Figure 8 shows some yield surfaces, obtained by means of equation (90), on Bishop's hydrostatic pressure-suction plane. Point A, which belongs to the initial yield surface, represents initial reservoir conditions. If for our calculations we use the compressibility modulus related to full saturation, which almost always happens in practice, the corresponding transformation is represented by the line A'-F'. For this path, which takes place with a compressibility modulus $\lambda(0)$, we have the same volumetric plastic strain as in the transformation A-E-F'' because the initial and final points of both loading paths belong to the same yield surfaces. The first part A-E involves only the reservoir pressure drop of 2 MPa with the compressibility modulus $\lambda(0.2)$ i.e. in partially saturated conditions, resulting in a volumetric plastic strain of 0.00274. The second transformation E-F'' represents the change in saturation with a volumetric plastic strain equal to 0.00282. The total plastic strain is hence 0.00556, the same as for A'-F'.

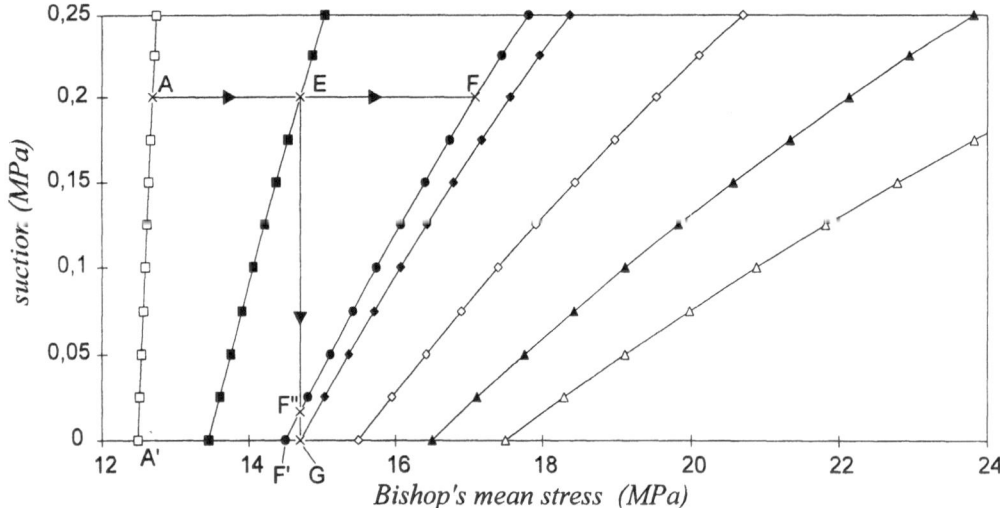

Figure 8. Yield surfaces in the Bishop's mean stress - suction plane (Simoni et al.).

The yield surface through F' could also be reached in partially saturated conditions by simply increasing the mean pressure change to 17.09 MPa and keeping the degree of saturation constant (point F). This is peculiar to partially saturated soil behaviour.

When analysing the total transformation A-E-F'' described, a possible error is the use of the compressibility modulus related to suction equal to 0.2 MPa, neglecting the effect of change in saturation. The error is dramatic (-51%). For a gas reservoir with the same thickness as the one previously analysed, assuming that fully saturated conditions can be reached after the change in reservoir pressure (point G), we obtain a compaction of 30 cm, whereas the additional compaction caused by pressure change is 22 cm.

We reach the conclusion that an increase in saturation after shutdown of the wells results in a considerable increment in reservoir compaction and resulting surface subsidence. This behaviour, which is typical of soils presenting a metastable structure, is a reasonable explanation for the subsidence recorded in some real cases after the closure of the reservoirs. Another consequence is that if the subsidence was predicted with compressibilities obtained from fully saturated samples, this subsidence presents an upper limit. On the contrary, in the case of metastable geomaterials, if compressibilities were used for the prediction of subsidence measured in partially saturated conditions (e.g. from markers downhole), this predicted subsidence is underestimated.

Finally, we apply the plasticity model which accounts for capillary effects in order to obtain the corrective term in time history of compaction of the gas reservoir which in Lewis and Schrefler (1998) was analysed assuming as compaction mechanism gas pressure drop only. To this end we use the time histories obtained by means of the traditional elastoplastic models as input data for gas pressure and cumulative water inflow (Lewis and Schrefler 1998). By using these data we determine the corresponding stress paths in (p', s) plane and calculate the ensuing volumetric plastic strains. This means that no particular numerical algorithm is necessary to integrate in time the constitutive equations. For the used numerical methodology we refer to Zienkiewicz et al. (1999). The time span considered covers forty years, during which water inflow, hence the change in saturation, is appreciable.

Figure 9 presents the time history for water saturation which is obtained by means of the inflowed water and the volume of gas remaining in the reservoir at depletion of the wells.

Figure 10 shows the history for gas pressure, which is similar to that of a real case (Evangelisti and Poggi 1970). In both Figures, point C represents the situation at the end of reservoir exploitation. Three other possible time histories are also depicted in the Figures. These are obtained by simplifying assumptions in the case of lack of a complete analysis of the transient problem. In particular the first corresponds to a linear variation between the extreme values of saturation and gas pressure (case a), the second does not account for gas pressure recovery after closure of the wells (case b) and the third (case c) assumes a linear variation of saturation with the same gas pressure history as in the real case. In all Figures the label real case represents the case study obtained using the data of the analyses in (Evangelisti and Poggi 1970, Lewis and Schrefler 1998).

The stress paths in the Bishop's mean stress-suction plane corresponding to the three situations are shown in Figure 11. It can be seen that a linear variation in (p', s) corresponds to a linear variation of relative saturation and gas pressure, whereas the almost vertical line C-B' corresponds to a constant gas value after the closure of the wells, which represent a transformation involving suction only.

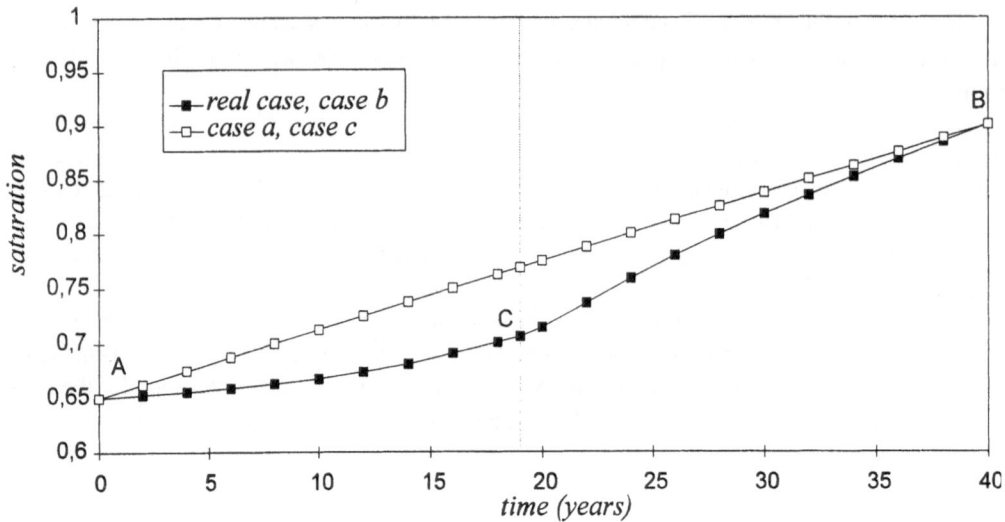

Figure 9. Saturation histories assumed for the elastoplastic analyses (Simoni et al.).

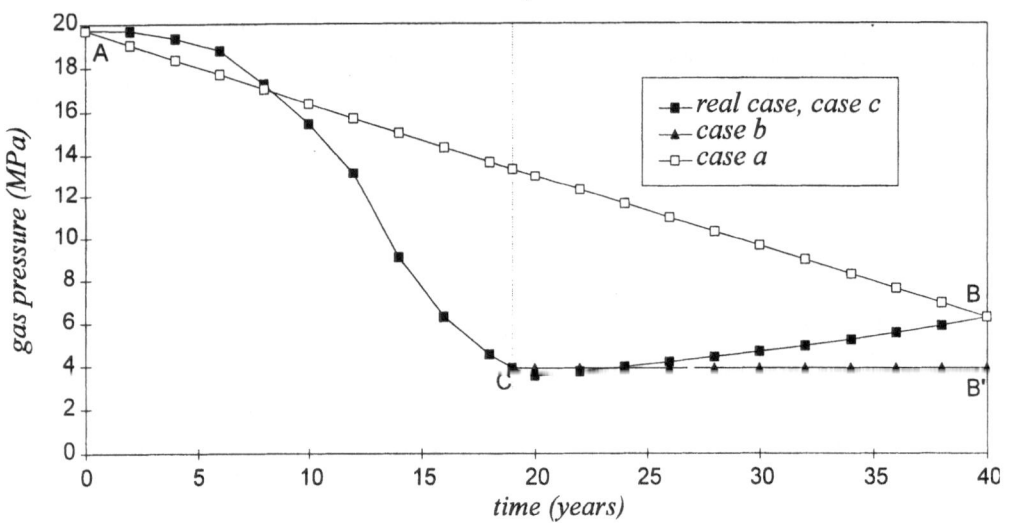

Figure 10. Gas pressure histories assumed for the elastoplastic analyses (Simoni et al.).

Figure 12 presents the results obtained showing the effect of capillary forces on reservoir compaction. Different compaction histories correspond to different stress paths. However, a common point between two displacement histories corresponds to a common point for the corresponding stress paths in the (p', s) plane. E.g. point C in Figure 12 for the real case and case b histories corresponds to point C in Figure 11.

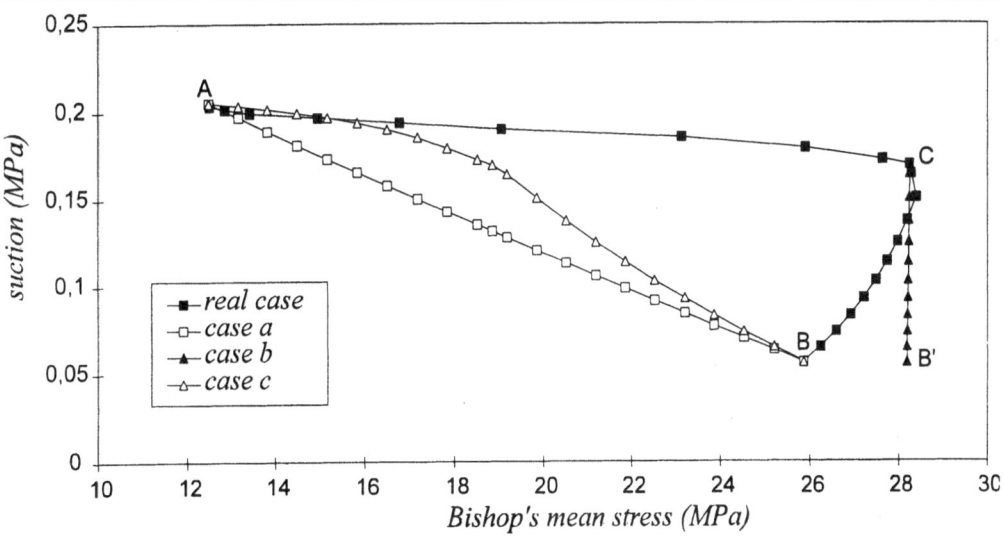

Figure 11. Stress paths in mean Bishop's stress-suction plane (Simoni et al.).

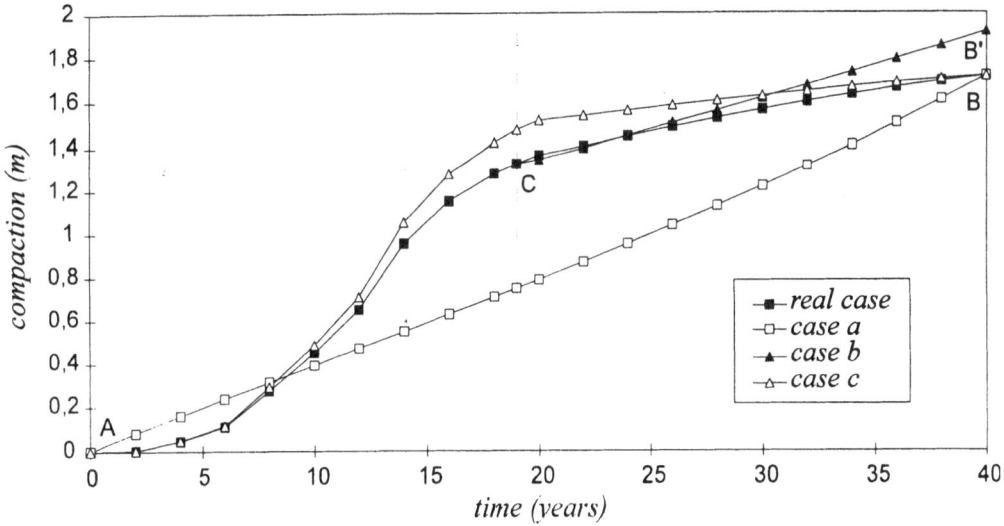

Figure 12. Calculated time histories for maximum reservoir compaction (Simoni et al.).

On the grounds of the results presented, some interesting remarks can be made:

compaction history strongly depends on the stress path followed;

in each case we obtain a compaction which increases after depletion of the wells. The adopted plasticity model is the only one which presents this feature, hence it can be applied to real problems to obtain believable numerical forecasts. In particular for the real case compaction rate decreases in time, as observed in reality;

recovery of pressure of the gas remaining in the reservoir limits the ongoing subsidence and its rate;

changes in saturation, at constant mean Bishop's pressure, result in volumetric plastic strain and remarkable vertical displacements.

7 Conclusions

A general theoretical framework has been presented together with its discrete solution to solve a rather broad class of environmental problems involving coupled thermo-hydro-mechanical processes and pollutant transport in geomaterials. The model is based on a strong physical background which allows clear identification of the constitutive equations needed to characterise the investigated medium. The governing equations have been obtained by means of volume averaging techniques and their discrete solution by means of the finite element method. The broad field of application confirms the generality of the presented framework.

References

Abriola L.M., Pinder G.F. (1985). A multiphase approach to modeling of porous media contamination by organic compounds 1. Numerical simulation, *Water Resources Research*, 21, 19-26.

AGIP (1996). Progetto Alto Adriatico-Studio di impatto ambientale, AGIP, San Donato, Italy.

Alonso E.E., Gens A. and Josa A. (1990). A constitutive model for partially saturated soils, *Géotechnique*, 40, 405-430.

Baggio P., Bonacina C., Schrefler B.A. (1997). Some considerations on modelling heat and mass transfer in porous media, *Transport in Porous Media*, 28, 233-281.

Bear J., and Bachmat Y. (1984). Transport phenomena in porous media – Basic equations, Fundamentals of Transport Phenomena in Porous Media. Bear, J. and Corapcioglu M.Y. eds. *Nato ASI Series*, Nijhoff.

Bolzon G., Schrefler B.A. and Zienkiewicz O.C. (1996). Elastoplastic soil constitutive laws generalized to partially saturated states. *Géotechnique*. 46, 279-289.

Brooks R.N., Corey A.T. (1966). Properties of Porous Media affecting fluid flow. *J. Irrig, Drain Div. Am.Soc.Civ.Eng.*, 92 (IR2). 61-68.

Corapcioglu M.Y., Baehr A.L. (1987). A compositional multiphase model for groundwater contamination by petroleum produces. 1. Theoretical considerations. *Water Resources Research*, 23. 191-200.

Daksanamurthy V. and Fredlund D.G. (1981). A mathematical model for predicting moisture flow in an unsaturated soil under hydraulic and temperature gradients. *Water Resources Research*, 17; 714-722.

Delage P., Schroeder C. and Cui Y.J. (1996). Subsidence and capillary effects in chalk. *Proceedings Eurock'96*, Balkema, Rotterdam, 1291-1298.

Evangelisti G. and Poggi B. (1970). Sopra i fenomeni di deformazione dei terreni da variazione di strato. *Atti Acc. delle Scienze dell'Istituto di Bologna*, Serie II, 6.

Gambolati G., Verri G. (1995). Advanced Methods for Groundwater Pollution Control. *CISM 364, Courses and Lectures*, Springer Verlag,.

Gray W.G, and Hassanizadeh S.M. (1991a). Paradoxes and realities in unsaturated flow theory, *Water Resources Research*, Vol. 27, 1847-1854.

Gray. W.G, and Hassanizadeh S.M. (1991b). Unsaturated flow theory including interfacial phenomena. *Water Resources Research*, Vol. 27, 1855-1863.

Hassanizadeh M. and Gray W.G (1979a). General conservation equations for multiphase systems: 1 Averaging procedure. *Advanced Water Resources*, 2, 131-144.

Hassanizadeh M. and Gray W.G (1979b). General conservation equations for multiphase systems: 2; Mass momenta, energy and entropy equations, *Advanced Water Resources*, 2, 191-203.

Hassanizadeh M. and Gray W.G (1980). General conservation equations for multiphase systems: 3. Constitutive theory for porous media flow, *Advanced Water Resources*, 3, 25-40.

Hassanizadeh M., and Gray W.G (1990). Mechanics and thermodynamics of multiphase flow in porous media including interphase transport, *Advanced Water Resources*, 13. 169-186.

Kueper B.H., Frind E.O. (1989). The behaviour of dense non-aqueous phase liquid contaminants in heterogeneous porous media. *Contaminant Transport on Groundwater*, Kobus H.E. and Kinzelbach W. eds., 381-387, Balkema, Rotterdam.

Lewis R.W. and Schrefler B.A. (1998). *The Finite Element Method in the Static and Dynamic Deformation and Consolidation of Porous Media*. Wiley and Sons, Chichester.

Lewis R.W., Schrefler B.A., Rahman N.A. (1998). A finite element analysis of multiphase immiscible flow in deforming porous media for subsurface systems. *Communications Numerical Methods Engineering*. 14, 135-149.

Manassero M., Shakelford C.D. (1994). The role of diffusion in contaminant migration through soil barriers. *Rivista Italiana di Geotecnica*, 1, 5-23.

Mendoza C.A., Frind E.O. (1990). Advective-dispersive transport of dense organic vapor in the unsaturated zone 1. Model development. *Water Resour. Res.*, 226, 379-387.

Meroi E., Schrefler B.A. (1995). Large strain static and dynamic hydro-mechanical analysis of porous media. From: Modern issues in Non-Saturated Soils, Chapter 9, A. Gens, P. Jouanna, Schrefler B.A., eds. *CISM Courses and Lectures 357* Springer Verlag, 397-447.

Olivella S., Gens A., Alonso E.E. and Carrera J. (1992). Constitutive modelling of porous salt aggregate. *Proceedings of the IV Int. Symposium On Numerical Models in Geomechanics.* A. A. Balkema, Roterdam, 179-189.

Park K.C., Felippa C.A. (1983). Partitioned analysis of coupled systems. Belytschko T. and Hughes T.R.J. eds. *Computational Methods for Transient Analysis.* Elsevier Science Publishers B.V.

Parker J.C. (1989). Multiphase flow and transportation in porous media, *Rev. Geoph.*, 27, 311-328.

Pastor M., Zienkiewicz O.C. and Chan A.H.C. (1990). Generalized plasticity and the modelling of soil behaviour. *Int. J. Numer. Anal. Methods Geomechanics.* 14, 151-190.

Schrefler B.A. (1995). F.E. in environmental engineering: Coupled Thermo-hydro-mechanical processes in porous media including pollutant transport. *Archives of Computational Methods in Engineering*, 2, 3, 1-54.

Schrefler B.A., Bolzon G, Salomoni V., Simoni L. (1997). On reservoir compaction in gas reservoirs. *Rendiconti Fis. Accademia Lincei*, s.9, v.8, 235-248.

Schrefler B.A., Zhan X.Y., Simoni L. (1995). A coupled model for water flow, airflow and heat flow in deformable porous media. *Int. J. Heat Flow and Fluid Flow*, 5, 531-547.

Schrelfler B.A., D'Alpaos L., Zhan X.Y. and Simoni L. (1994). Pollutant transport in deforming porous media, *Eur. J. Mech., A/Solids*, 13, n. 4-suppl., 175-194.

Simoni L., Salomoni V. and B. A. Schrefler. Elastoplastic Subsidence Models with and without Capillary Effects. *Computer Methods in Applied Mechanics and Engineering*, in print.

Turska E., Schrefler B.A. (1993). On convergence condition of partitioned solution procedures for consolidation problems, *Computer Methods in Applied Mechanics and Engineering*, 106, 51-63.

Zhan X., Schrefler B.A., Simoni L. (1995). Finite element simulation of multiphase flow, heat flow and solute transport in deformable porous media. *Proceedings International Conference Finite Elements in Fluids*, Morandi Cecchi M., Morgan K., Periaux J., Schrefler B.A., Zienkiewicz O.C. eds. Dipartimento di Matematica Pura ed Applicata, University of Padua, part II, 1283-1290.

Zienkiewicz O.C. and Taylor R.L. (1989), *The Finite Element Method*, McGraw-Hill, London.

Zienkiewicz O.C., A. Chan, M. Pastor, Schrefler B.A. and T. Shiomi (1999). *Computational Soil Dynamics with special Reference to Earthquake Engineering.* J. Wiley, Chichester.

Appendix

$$\mathbf{K}_T = -\int_\Omega \mathbf{B}^T \mathbf{C}_T \mathbf{B} d\Omega$$

$$\mathbf{C}_{sw} = \int_\Omega \mathbf{B}^T \overline{\alpha}\, \mathbf{m}\left[S_w + \left(p^g - p^w\right)\frac{\partial S_w}{\partial p^c}\right] \mathbf{N} d\Omega$$

$$\mathbf{C}_{sa} = \int_\Omega \mathbf{B}^T \overline{\alpha}\, \mathbf{m}\left[S_g - \left(p^g - p^w\right)\frac{\partial S_w}{\partial p^c}\right] \mathbf{N} d\Omega$$

$$\mathbf{C}_{st} = \int_\Omega \mathbf{B}^T \left[\frac{\beta_s}{3}\mathbf{D}_T \mathbf{m} - \overline{\alpha}\left(p^g - p^w\right)\frac{\partial S_w}{\partial T}\mathbf{m}\right] \mathbf{N} d\Omega$$

$$\dot{\mathbf{f}}_s = \int_\Omega \mathbf{N}^T \frac{d\mathbf{b}}{dt} d\Omega + \int_\Gamma \mathbf{N}^T \frac{d\mathbf{t}}{dt} d\Gamma$$

$$\mathbf{C}_{ws} = \int_\Omega \mathbf{N}^T \overline{\alpha} S_w\, \mathbf{m}^T\, \mathbf{B}\, d\Omega$$

$$\mathbf{P}_{ww} = \int_\Omega \mathbf{N}^T \left\{ -n\frac{\partial S_w}{\partial p^c} + \frac{nS_w}{K_w} + S_w\frac{\overline{\alpha} - n}{K_s}\left[S_w + \left(p^g - p^w\right)\frac{\partial S_w}{\partial p^c}\right]\right\} \mathbf{N} d\Omega$$

$$\mathbf{C}_{wa} = \int_\Omega \mathbf{N}^T \left\{ n\frac{\partial S_w}{\partial p^c} + S_w\frac{\overline{\alpha} - n}{K_s}\left[S_g - \left(p^g - p^w\right)\frac{\partial S_w}{\partial p^c}\right]\right\} \mathbf{N} d\Omega$$

$$\mathbf{H}_{ww} = \int_\Omega \left(\nabla \mathbf{N}\right)^T \mathbf{k}^w \nabla \mathbf{N} d\Omega \qquad\qquad \mathbf{k}^w = \frac{kk^{rw}\rho^w g}{\mu^w}$$

$$\mathbf{C}_{wt} = \int_\Omega \mathbf{N}^T \left\{ \left[n - \frac{\left(\overline{\alpha} - n\right)S_w}{K_s}\left(p^g - p^w\right)\right]\frac{\partial S_w}{\partial T} - \left[n\beta_w + \left(\overline{\alpha} - n\right)\beta_s\right]S_w\right\} \mathbf{N} d\Omega$$

$$\dot{\mathbf{f}}_w = \int_\Omega \left(\nabla \mathbf{N}\right)^T \mathbf{k}^w \rho_w g d\Omega - \int_\Gamma \mathbf{N}^T q^w d\Gamma$$

$$\mathbf{C}_{as} = \int_\Omega \mathbf{N}^T \overline{\alpha} S_g\, \mathbf{m}^T\, \mathbf{B}\, d\Omega$$

$$\mathbf{C}_{aw} = \int_\Omega \mathbf{N}^T \left\{ n\frac{\partial S_w}{\partial p^c} + S_g\frac{\overline{\alpha} - n}{K_s}\left[S_w + \left(p^g - p^w\right)\frac{\partial S_w}{\partial p^c}\right]\right\} \mathbf{N} d\Omega$$

$$\mathbf{P}_{aa} = \int_\Omega \mathbf{N}^T \left\{ -n\frac{\partial S_w}{\partial p^c} + \frac{nS_g}{K_g} + S_g\frac{\overline{\alpha} - n}{K_s}\left[S_g - \left(p^g - p^w\right)\frac{\partial S_w}{\partial p^c}\right]\right\} \mathbf{N} d\Omega$$

$$\mathbf{H}_{aa} = \int_\Omega \left(\nabla \mathbf{N}\right)^T \mathbf{k}^g \nabla \mathbf{N} d\Omega \qquad\qquad \mathbf{k}^g = \frac{kk^{rg}\rho^g g}{\mu^g}$$

$$\mathbf{C}_{at} = \int_{\Omega} \mathbf{N}^T \left\{ \left[n + \frac{(\bar{\alpha}-n)S_g}{K_s}(p^g - p^w) \right] \frac{\partial S_w}{\partial T} - \left[n\beta_g + (\bar{\alpha}-n)\beta_s \right] S_g \right\} \mathbf{N}d\Omega$$

$$\dot{\mathbf{f}}_a = \int_{\Omega} (\nabla \mathbf{N})^T \mathbf{k}^g \rho_g \mathbf{g} d\Omega - \int \mathbf{N}^T q^g d\Gamma$$

$$\mathbf{C}_{tw} = \int_{\Omega} \mathbf{N}^T \left[\frac{(1-n)\rho_s C_s}{K_s} \left(S_w + (p^g - p^w)\frac{\partial S_w}{\partial p^c} \right) - n\rho_w C_w \frac{\partial S_w}{\partial p^c} \right.$$
$$\left. + \frac{nS_w \rho_w C_w}{K_w} + n\rho_g C_g \frac{\partial S_w}{\partial p^c} \right] T\mathbf{N}d\Omega$$

$$\mathbf{C}_{ta} = \int_{\Omega} \mathbf{N}^T \left[\frac{(1-n)\rho_s C_s}{K_s} \left(S_g - (p^g - p^w)\frac{\partial S_w}{\partial p^c} \right) + n\rho_w C_w \frac{\partial S_w}{\partial p^c} \right.$$
$$\left. + \frac{nS_g \rho_g C_g}{K_g} - n\rho_g C_g \frac{\partial S_w}{\partial p^c} \right] T\mathbf{N}d\Omega$$

$$\mathbf{P}_{tt} = \int_{\Omega} \mathbf{N}^T \left\{ \left[\frac{(n-1)\rho_s C_s}{K_s}(p^g - p^w)\frac{\partial S_w}{\partial T} + (n-1)\beta_s \rho_s C_s + n\rho_w C_w \frac{\partial S_w}{\partial T} \right. \right.$$
$$\left. - nS_w \beta_w \rho_w C_w - n\rho_g C_g \frac{\partial S_w}{\partial T} - n(1-S_w)\beta_g \rho_g C_g \right] T$$
$$\left. + (1-n)\rho_s C_s + nS_w \rho_w C_w + n(1-S_w)\rho_g C_g \right\} \mathbf{N}d\Omega$$

$$\mathbf{H}_{tt} = \int_{\Omega} \mathbf{N}^T \left(S_w \rho_w C_w \mathbf{k}^w p_{,i}^w + S_g \rho_g C_g \mathbf{k}^g p_{,i}^g \right) \nabla \mathbf{N}d\Omega + \int_{\Omega} (\nabla \mathbf{N})^T \lambda \nabla \mathbf{N}d\Omega$$

$$\dot{\mathbf{f}}_t = \int_{\Omega} \mathbf{N} \left[(1-n)\rho_s Q_s + n\rho_w S_w Q_w + n\rho_g S_g Q_g \right] d\Omega - \int_{\Gamma} \mathbf{N}q^T d\Gamma$$

$$\mathbf{C}_{wc_w} - \int_{\Omega} \mathbf{N}^T \left[-\frac{(\bar{\alpha}-n)S_w}{K_s}(p^g - p^w)\frac{\partial S_w}{\partial c_w^\pi} + n\frac{\partial S_w}{\partial c_w^\pi} + nS_w \gamma_w \right] \mathbf{N}d\Omega$$

$$\mathbf{C}_{wc_a} = \int_{\Omega} \mathbf{N}^T \left[-\frac{(\bar{\alpha}-n)S_w}{K_s}(p^g - p^w)\frac{\partial S_w}{\partial c_a^\pi} + n\frac{\partial S_w}{\partial c_a^\pi} \right] \mathbf{N}d\Omega$$

$$\mathbf{C}_{ac_w} = \int_{\Omega} \mathbf{N}^T \left[-\frac{(\bar{\alpha}-n)S_g}{K_s}(p^g - p^w)\frac{\partial S_w}{\partial c_w^\pi} - n\frac{\partial S_w}{\partial c_w^\pi} \right] \mathbf{N}d\Omega$$

$$\mathbf{C}_{ac_a} = \int_{\Omega} \mathbf{N}^T \left[-\frac{(\bar{\alpha}-n)S_w}{K_s}(p^g - p^w)\frac{\partial S_w}{\partial c_a^\pi} - n\frac{\partial S_w}{\partial c_a^\pi} + nS_g \gamma_g \right] \mathbf{N}d\Omega$$

$$\mathbf{C}_{tc_w} = \int_{\Omega} \mathbf{N}^T \left[-\frac{(1-n)\rho_s C_s}{K_s}(p^g - p^w)\frac{\partial S_w}{\partial c_w^\pi} + n\rho_w C_w \frac{\partial S_w}{\partial c_w^\pi} - n\rho_g C_g \frac{\partial S_w}{\partial c_w^\pi} + n\rho_w S_w C_w \gamma_w \right] T\mathbf{N}d\Omega$$

$$\mathbf{C}_{tc_a} = \int_\Omega \mathbf{N}^T \left[-\frac{(1-n)\rho_s C_s}{K_s}\left(p^g - p^w\right)\frac{\partial S_w}{\partial c_a^\pi} + n\rho_w C_w \frac{\partial S_w}{\partial c_a^\pi} - n\rho_g C_g \frac{\partial S_w}{\partial c_a^\pi} + n\rho_g S_g C_g \gamma_g \right] T \mathbf{N} d\Omega$$

$$\mathbf{C}_{c_w w} = \int_\Omega \mathbf{N}^T \left[-n\frac{\partial S_w}{\partial p^c} \right] c_w^\pi \mathbf{N} d\Omega$$

$$\mathbf{C}_{c_w t} = \int_\Omega \mathbf{N}^T \left[n\frac{\partial S_w}{\partial T} \right] c_w^\pi \mathbf{N} d\Omega$$

$$\mathbf{C}_{c_w a} = \int_\Omega \mathbf{N}^T \left[n\frac{\partial S_w}{\partial p^c} \right] c_w^\pi \mathbf{N} d\Omega$$

$$\mathbf{P}_{c_w c_w} = \int_\Omega \mathbf{N}^T \left[n\frac{\partial S_w}{\partial c_w^\pi} c_w^\pi + nS_w \right] \mathbf{N} d\Omega$$

$$\mathbf{C}_{c_w c_a} = \int_\Omega \mathbf{N}^T \left[n\frac{\partial S_w}{\partial c_a^\pi} \right] c_w^\pi \mathbf{N} d\Omega$$

$$\dot{\mathbf{f}}_{c_w} = \int_\Omega \mathbf{N}\left[I_w + \Gamma^w \right] d\Omega - \int_\Gamma \mathbf{N} q^{c_w} d\Gamma$$

$$\mathbf{H}_{c_w c_w} = \int_\Omega \mathbf{N}^T \mathbf{k}^w p_{,i}^w \nabla \mathbf{N} d\Omega + \int_\Omega (\nabla \mathbf{N})^T nS_w \mathbf{D}_w^\pi \nabla \mathbf{N} d\Omega$$

$$\mathbf{C}_{c_a w} = \int_\Omega \mathbf{N}^T \left[n\frac{\partial S_w}{\partial p^c} \right] c_a^\pi d\Omega$$

$$\mathbf{C}_{c_a a} = \int_\Omega \mathbf{N}^T \left[-n\frac{\partial S_w}{\partial p^c} \right] c_a^\pi d\Omega$$

$$\mathbf{C}_{c_a t} = \int_\Omega \mathbf{N}^T \left[-n\frac{\partial S_w}{\partial T} \right] c_a^\pi d\Omega$$

$$\mathbf{C}_{c_a c_w} = \int_\Omega \mathbf{N}^T \left[-\frac{\partial S_w}{\partial c_w^\pi} \right] c_a^\pi d\Omega$$

$$\mathbf{P}_{c_a c_a} = \int_\Omega \mathbf{N}^T \left[n\frac{\partial S_w}{\partial c_a^\pi} c_a^\pi + nS_g \right] \mathbf{N} d\Omega$$

$$\dot{\mathbf{f}}_{c_a} = \int_\Omega \mathbf{N}\left[I_a + \Gamma^a \right] d\Omega - \int_\Gamma \mathbf{N} q^{c_a} d\Gamma$$

$$\mathbf{H}_{c_a c_a} = \int_\Omega \mathbf{N}^T \mathbf{k}^a p_{,i}^a \nabla \mathbf{N} d\Omega + \int_\Omega (\nabla \mathbf{N})^T nS_g \mathbf{D}_g^\pi \nabla \mathbf{N} d\Omega$$